Love and Other Technologies

Love and Other Technologies

Retrofitting Eros for the Information Age

Dominic Pettman

FORDHAM UNIVERSITY PRESS
NEW YORK 2006

Library of Congress Cataloging-in-Publication Data

Pettman, Dominic.
Love and other technologies : retrofitting eros for the information age / Dominic Pettman.—1st ed.
p. cm.
Includes bibliographical references and index.
ISBN-13: 978-0-832-2668-9 (cloth : alk. paper)
ISBN-10: 8232-2668-9 (cloth : alk. paper)
ISBN-13: 978-08232-2669-6 (pbk. : alk. paper)
ISBN-10: 0-8232-2669-7 (pbk. : alk. paper)
1. Love. 2. Technology. 3. Civilization, Modern—21st century. I. Title.
BD436.P425 2006
128′.46—dc22

2006032362

Printed in the United States of America
08 07 06 5 4 3 2 1
First edition

CONTENTS

ACKNOWLEDGMENTS

This book came in fits and spurts over a six-year period, from 2000 to early 2006. During this time I taught in several different environments and institutions; and in a sense, this project was one of the only tangible elements which gave a sense of continuity to my daily life, threading its *problématique* through the streets of Melbourne, Geneva, Amsterdam, and New York. I am not in a position to tell whether these different places—and more importantly, the people I encountered—have any noticeable influence on different passages or chapters. What I *do* know for certain is that it is impossible to rethink the concepts of love, community, and technology without exposing oneself to nearly lethal doses of each. And for this reason, I'd like to acknowledge those who wittingly, unwittingly, or even despite their express wishes had an impact on the following pages.

Wlad Godzich, Rick Waswo, Thomas Elsaesser, Simon During, Steven Shaviro, Carl Skelton, Sumita Chakravarty, and McKenzie Wark acted as patrons, benefactors, mentors, coconspirators, and intellectual inspiration. Pierre Grosjean, Drehli Robnik, Gabu Heindl, Wanda Strauven, Ria Thanouli, and Malte Hagener provided vodka-infused companionship and lucid arguments to my more half-baked assertions. Justin Clemens, Karolina Krebs, Eddie Maloney, Dan Ross, David Odell, and Carrie Olivia Adams all deserve special mentions for the very different ways in which they embody the more utopian possibilities which lie latent in each moment (and because they email me a lot).

An earlier version of the Introduction was published in *Parachute*, no. 101, 2001. This research was first undertaken in the English with Cultural Studies Department, University of Melbourne, where I am currently an Honorary Fellow.

My endless gratitude goes to Helen Tartar for her long-term interest and faith in my manuscript, as well as the anonymous readers, whose sharp-eyed and illuminating feedback helped refine the final product.

Finally, my greatest thanks go to Merritt, who on a daily basis reminds me why love is such a crucial concept for us all to face head-on.

PREFACE: A NEW KIND OF LOVE

In Hermann Broch's 1931 novel, *The Anarchist*, the protagonist, August Esch, wanders through the hallowed halls of the head offices of the Central Rhine Shipping Company, having recently accepted a job there as an accountant. He stops short upon reading a woman's name on one of the seemingly endless office doors and

> felt a sudden desire for the unknown woman behind the door, and there arose in him the conception of a new kind of love, a simple, one might almost say a business-like and official kind of love, a love that would run as smoothly, as calmly, and yet as spaciously and never-endingly, as these corridors with their polished linoleum. (59)

This "new kind of love" is not only hatched in the heart of Esch but linked to a world-historical condition via an experience we could dub, with a nod to Kafka, the bureaucratic sublime. It is a cold love. A smooth love. Balanced in the same fashion, and according to the same principles, as company accounts.

As Broch makes clear, in some obscure way this new kind of love is linked to gender and can quickly switch into a different mode:

> But then he saw the long series of doors with men's names, and he could not help thinking that a lone woman in that masculine environment must be as disgusted with it as Mother Hentjen was with her business. A hatred of commercial methods stirred again within him, hatred of an organization that, behind its apparent orderliness, its smooth corridors, its smooth and flawless book-keeping, concealed all manner of infamies. And that was called respectability! (Ibid.)

And so, on the one hand, we have that ancient principle "love," and on the other, we have a new genus, perhaps a genetically modified strain, developed in order to survive modern economic exigencies. Such a new species of love has different names according to different agendas. Michael Hardt and Antonio Negri call it "affective labor," while Zygmunt Bauman calls it "liquid love" (a symptom of so-called liquid modernity). But whichever name this mutation of feeling masquerades under, it prompts us to examine the mysterious relationship between longing and *be*-longing in an era often characterized as isolated, discordant, bankrupt, godforsaken, and even inhuman.

Within this rhetorical context, it is worth speculating whether love is the only discourse still available to us that is capable of salvaging singularity in a late capitalist epoch, or whether it is rather a case that "love" has become (or perhaps always was) a decoy that lures us into a libidinal economy no less indifferent to individual suffering than the macroeconomy overseen by the IMF and the World Bank.

Love today is performed and assessed not so much in an age of cholera as one of BSE and CJD, pathogens created for us by a cannibalistic regime of "mechanically recovered meat" (a technical term from the abattoir industry but equally appropriate for almost all other kinds of commerce, of which the dating and entertainment industries are only the most obvious). Hence, Stendhal's famous definition of love as a "crystallization" has taken on a sinister meaning in a time when one bite of a hamburger can lead to a very literal crystallization of brain tissue (1975).

Several questions emerge from such ambient conditions. Is love a fetish? Is our emotional surrender to the phrase "I love you" a disavowal, in the Freudian sense, or a delusion, in the Marxist sense? Is this institutionalized form of desire the addictive element which is actually poisoning the fragile ego-system of twenty-first-century social life?

And yet, I do not simply want to regurgitate that particular strain of Continental philosophy which can be boiled down to the statement "Things may not have been better before, but they are certainly getting worse." There is sensible work to be done, sleeves to be rolled up, paradigms to be dismantled.

For instance, we often ponder the meaning supporting the Ur-sentence "I love you." Remaining faithful to a kind of Howard-Jones effect, we have concentrated most of our efforts on the word "love," rather than questioning the "I" and the "you," as if the bridge were more important than the banks it claims to connect. Indeed, we tend to assume that this metaphor of bridging two stable and identifiable entities or locations is enough to capture the passionate event, which it clearly is not. Love is more than merely an engineering feat which joins two previously isolated points. (Remember, no man is an island.) In contrast, love (perhaps I can drop the quotation marks for this word at this point, or at least render them invisible) must negotiate the "being singular plural" (Nancy) of beings. It must recognize the fact that "we" are not completely coincident and/or consistent with ourselves. That is to say, our imagined identities never fully match that bundle of behaviors which comport themselves under our name. To put it yet another way, our shadows (fantasies, delusions, self-projected images) are of different shapes from those which cast the shadows (not to be mistaken with a "true-self," but closer to a heterogeneous set of elements revolving around an absent organizing center).[1]

The significance of this approach to subjectivity will become more apparent as my argument unfolds, but at this point it is necessary to establish the flexible idea of ontological overlaps, rather than the rigid metaphysical buffer zones of Cartesian intersubjectivity (the reasons I don't say "Platonic intersubjectivity" are explained in chapter 1). We live so much through the media that it seems almost tautological to say that the world is mediated. When the medium becomes the message, and vice versa, it is perhaps time to start talking about immediacy again. And it is in this context that love enters the equation.

For example, think of those everyday ex-pathic circuits which trace the Möbius strip of "me" versus "not-me." We say "Oops" when someone is dropping a carton of eggs.[2] We yawn when we see someone else doing the same. Such contagion is not necessarily a case of empathy, the human capacity to imagine oneself in another's place, but a way for *community* or *technology* or *intelligence* to manifest and transmit itself. From this perspective a yawn is a cosmic entity which enlists people as its host in order to bring itself into being. And who is to say that this is not a valid perspective (especially when we replace the seemingly trivial example of "yawn" with, say, "music" or "art" or "culture itself")?

"Here's Looking at You, Kid"

"Separate the sexual act from love, and the language of love is devalued." So states Anthony Burgess (1978, 102). In the very next sentence, he goes on to say, "An aspect of our freedom is our right to debase the language totally, so that its syntagms become mere noise." The link between the first and the second sentence is left intriguingly open, inviting us to fill in the blanks with our own thoughts on the relationship between eros, agapē, communication, and structure (and not only that, but to reflect on the notion of "relationship" itself).

Identifying the signal-to-noise relationship of amorous discourse seems like a technical imperative in these troubled times.[3] While at first glance a study of love may seem like a retreat into the personal, it will soon become apparent that it is anything but. Love, like weaponized anthrax spores, has a habit of getting into everything.

One entry point we can use for such an all-encompassing and diffuse discourse is the visual mechanism which often triggers its existence in the first place.[4] As Kaja Silverman maintains, "Passion is a semiotic affair" (2000, 50). Indeed, this same critic identifies a formal transhistorical human subject, the "world spectator," who is referred to as such because they are *subject* to the scopic regime of love: "The world spectator is emphatically a desiring subject" (11). Describing a conceptual trajectory linking Plato's hypnotized cave dwellers with Debord's equally mesmerized cinemagoers, Silverman focuses on the ontological feedback loops of (visual) attraction:

> Were others to look at us through our own eyes, 'ourselves' is precisely what we would never be. We can appear, and so Be, only if others 'light' us up. To be lit up means to be seen from a vantage point from which we can never see ourselves. It also means to embody not our own, but *someone else's* idea of beauty. Our 'essence' is thus strangely nonessential. . . . The being whom I light up with the radiance of affirmation supplies me with the form which allows me to see what I could not otherwise see; it alone makes possible beauty's embodiment. It is together, then, that we bring about its appearance. (19–21)

If Beauty goes for a walk alone in the forest, would Beauty still be beautiful? The answer is an emphatic No.

Thus, for Silverman—as for myself—"libidinal signification has an *ontological* force" (43). That is to say, every relationship is a transductive relationship, one which doesn't merely link but *creates* the terms in that same relationship. Beauty—or more simply Self, whether figured as beautiful or ugly or somewhere in between—does not precede the encounter with Others. This "ontological force," the one emphasized throughout the following chapters, prompts an ethical assessment of one's own actions, not so much in the moralistic-behavioral sense ("Am I a good person?") but according to a more symbolic economy of recognition and deferral ("Am I a person?"). Silverman states: "To each of us, through our particular libidinal history, has been given the potentiality for participating in a unique series of disclosures. This potentiality is not so much a talent as a responsibility. When we fail to realize it, we are bottomlessly guilty" (48).

Without embracing this statement completely (in what sense is this series *unique*, and why give so much ground to the deployment of guilt; do we detect a whiff of emotional blackmail here?), the notion of responsibility in relation to an erotics of Being is nevertheless a revealing one. Through it we can *see* the stakes involved in being together, longing together, and belonging together (in the sense of W. J. T. Mitchell's desire to "make seeing show itself" [2002, 175]).

For in English at least, the verb "to look" can be both objective ("You look good"), and subjective ("I look at you"). It is simultaneously passive and active, gazing and gazed upon. And yet, from the myth of Aphrodite's lazy eye to those infamous blinkers which fall over the retinas of the impassioned, love is an interactive event which involves not only sight but all of the senses—and perhaps others we are not even aware of . . . yet.

To coin a phrase, then: If love is blind, sex is Braille.

Techtonic Movements

Love, technology, community.

It is my suggestion in this book that these three terms in fact designate the same thing, or at least the same movement—specifically, a movement toward the other.

If we recall the bone-wielding man-ape in Stanley Kubrick's *2001: A Space Odyssey* (1968), we have access to a fable of the origin of tools—tools which begin under the sign of weaponry and hostility. The primate, known as Moon-Watcher, beating another simian to death with a bone, certainly fails to qualify as an originary figure of love or community. Yet, when viewed as a negative template, such an image provides the outlines of an encounter which can allow us to think through the emergence and legacy of these crucial cultural and political terms. That is to say, if we suspend for a moment the idea of motivation, human social life can be figured as always already a movement toward others. Whether this movement is born of violence or tenderness, the vector remains the same. Simply put, we would do well at the outset to avoid dividing "love" on one side of the fence and "hatred" on the other—and the same can be said for "technology" and "nature," or "community" and "alienation."[5]

For instance, definitive distinctions between tools and weapons are notoriously difficult to sustain, as the case of the spanner, which can be used to tighten a bolt or seize the machinery itself. And as the West shifts further into a postmillennial paranoid phase in relation to the Scylla and Charybdis of "terror" and "security," the emphasis on so-called technological superiority only intensifies. As theorists such as Manuel de Landa, Paul Virilio, and Friedrich Kittler have shown, cutting-edge technologies are produced and developed in the context of the incessant war machine, then later trickle down into other spheres, such as medicine, civil engineering, agriculture, and, eventually, domestic convenience.

Despite this ancient alliance between technology and violence, itself generated by the engines of alterity which are used to distinguish allies from enemies, there is an argument to be made that *eros*, *techne*, and *communitas* are so inextricably linked as to often be indistinguishable. Just as love can spasm simultaneously into hate, the words we use to designate these phenomena cannot adequately capture the dynamic relationships which comprise the immediacy of the moment and the enigmas of history (or indeed the enigmas of the moment and the immediacy of history). It is the triangulation of these terms that reveals some of the blind spots in our current discussions and debates of the role of technology in contemporary life.

At the end of my previous book, *After the Orgy* (2002), I signaled a "preemptive" exhaustion back on the other side of the year 2000, the already inscribed expectation of a kind of depleted politics. This phenomenon, I argued, both informs and creates the liminal space between anticipation and anticlimax, a space which need not necessarily be as debilitating as it appears in daily life. In this sequel of sorts, I explore the ethical implications of this posthuman(ist) politics in relation to contemporary media and literature. This book thus traces the various vectors and movements of emergent globalization, specifically in order to locate and identify the cultural stakes which are forged at the rhetorical intersections between love, technology, and community.

While my previous project relied heavily on Freud, Nietzsche, Baudrillard, and Bataille, this new book primarily relies on more recent, poststructuralist theories of communication; but with a twist. In place of those posed by the elongated Ur-concepts of first-wave psychoanalysis and transgressive philosophy, my methodology is both pragmatic and strategic, enlisting ideas poached from different contexts and epochs to support an argument which witnesses change over time and yet maintains a certain formal consistency. (Love is always already a technology of being together, yet the experience of "being" and "together" is something composed in historical time.)

Michel Foucault identifies the issue when he states that certain codes can remain remarkably stable and consistent through different periods, while the ethics surrounding them change by way of integration, displacement, and reutilization. Ruptures occur in different historical contexts, reconfiguring the symbolic systems by which we are constituted and through which we articulate new phases and techniques of symbolization. But as Foucault reminds us, "There is a technology of the constitution of the self which cuts across symbolic systems" (1988, 369).

My approach thus takes a leaf from this Foucaultian fracture, and strategically emphasizes continuity when it is significant, and discontinuity when it tangibly impacts or disrupts codification (in my case, of intimacy). Indeed, part of the difficulty with this project is my own desire to sidestep the lumbering linearity of diachronic analysis. One method is to superimpose symptomatic synchronicities (a fancy way of saying "to compare and contrast different epochs") in order to circumvent temporally discrete periods. This calls for an analytic which mimics Hitchcock's famous camera technique in *Vertigo* (1958): pulling the camera back while zooming in at the same time. Periodization is approached as both an open problem and a taxonomic technology in its own right.

From a certain angle, to trace the influence of technology through time is to beg this question of periodization. However, it is clear that chronological models are failing us when it comes to understanding something this "patent and deranging." For Heidegger, the question of technics is intimately bound up with one's quotidian-epistemic wherewithal. In his scheme, both techne and episteme "mean to be entirely at home in something, to understand and be expert in it" (1977a, 13). Or at least it *did* mean this. Before "the turning," world-historical periods could be marked by the age before and after the "world-picture," or the increase in "unreasonable demands" placed upon nature. And yet for those feeling decidedly unhomed, or homeless, in the digital age, technics obliterates our access to such neat, even quaint concepts as "before and after," by virtue (if that is the right word) of its own momentum, or blind becoming.[6]

Hence my decision to drop the original subtitle of this book, *New Questions for Old Media, Old Questions for New Media*, both for being unwieldy and for its inability to recognize the undecidability of technics in relation to origins, overlappings, applications, innovations, and mutations. By complicating the relationship of past to present to future, precisely by suspending it, is to go beyond—or at least seek a space apart from—repetition-as-difference (for instance, McLuhan's vicious spiral of tribalism to modernity to neotribalism).

No doubt, with a canvas this vast, it is prudent to give some of the more dominant names in the field a certain amount of shore leave. Those listed above, along with seemingly obvious candidates like Derrida and Lacan, thus make only cameo appearances in the following pages (although it would be a mistake to underestimate their influence). Instead, Jean-Luc Nancy, Bernard Stiegler, Gilles Deleuze, and Giorgio Agamben act as our primary guides to the material, since they function as my own technologies for cutting through the Gordian Knot of mediated desire.[7]

It is these thinkers who taught me that definitions of *technology* should not be limited to industrial infrastructure and other "machines that go beep," but should extend to specific relationships involving power, knowledge, and discourse (often based on the biomechanics of natural organisms and processes). It is they who emphasize the facts that language itself can be considered a technology and that the very concept of "the human" is being reinvented by rhetorical and technological drift. They offer hope in a climate where the invocation of the very word is often enough to confirm that there is none. What's more, as with all cultural phenomena, there are periods of waxing and waning. Some critics and concepts seem to bear a heavy burden of overcitation and déjà vu, while others build on these in more timely and less familiar ways. (My last book was deliberately untimely; this one attempts to be pre-timely, or "ahead of the curve," as the corporate analysts say.)

Yet no useful idea is rejected merely because it happens to come from a dubious or unfashionable source. Kierkegaard is connected to Bergson, Luhmann to Silverman, and Barthes to Kittler in a concerted effort to "think diagonally," cutting across the artificial fences which are still—even in the age of Monsanto—built and mended, then rebuilt and remended, in order to keep the trespassers out of disciplinary fields.

The movement I wish to trace is, for my purposes at least, a *techtonic* movement. The question then becomes: What logic informs this particular neologism? Well, specifically the logic of the planet itself, which is not only hurtling around a finite star but also shifting internally according to its own geological imperatives, tectonic forces bringing buried elements up from the molten mantle and plunging vast continental plates back downward on enormous lithic conveyer belts. The texts, spaces, people, practices, and places documented in this book are framed in terms of parallel "techtonic movements," those being the constantly shifting plates and planes which make up the dubious foundations of our cultural assumptions. As with the earth itself, these movements can be slow and imperceptible, indifferent to the vicissitudes of human time, or else they can be quick and violent, as occurs with volcanic eruptions and earthquakes.[8]

Changing the spelling of *tectonics to techtonics*, however, obviously emphasizes the technological element of these movements and therefore subscribes to an etiology of "the human" largely indebted to Bernard Stiegler, as well as to the various intellectual giants on whose dandruff-covered shoulders he so brilliantly perches. It thus becomes necessary to trace an emergent globalism in order to locate and identify the ultimately political stakes which are forged at the textual intersections between love, technology, and community.

Such an approach necessarily draws us into a wider constellation of more general questions regarding passion and legibility, essentialism and rhizomatics, historical sequentiality and its construction, nostalgia and its reconstruction, and that network of discourses which articulate social behavior within the mediascape of the electro-information era. In order to map the movement of these techtonic fluxes, we must therefore consider existing definitions of cultural transmission, difference, representation, moralized monogamy, and their potential alternatives. The hope is that in thinking such notions through, we can unearth adequate evidence to counter the overdetermined geopolitical axes of our age, on both micro and macro levels.

Perhaps this project springs from my own interest in a "politics of exhaustion," an orientation suspicious of those movements which seek to harness energy and optimism, as if these forces were not constantly vulnerable to being funneled into deep-cut circuits of Capital, Art, Expression, Charity, Street Theater, and other horrors of the modern world. Its intention is not to fetishize or romanticize the Other but rather to listen attentively to the feedback loop produced when "self" and "other" give way to an ethical assemblage beyond intersubjectivity.

As David Odell points out in his online book *A Rushed Quality:*

> It is remarkable that we can forget our primordial sociality to such a degree that the "constitution of intersubjectivity" seems to be a legitimate question. What is the evidence for basal solipsism? We die alone, we fear others, we can never know what they are thinking much less understand them, we must study to bend them to our will and so on. As well to say that all of these must be constituted on prior sociality, and far easier to sketch out the lines. What an absurd imposture the solitary self is! It is a myth that gives consistency to our emotions, our desires and our ignorance. And to our loves as well? (2001)

Introduction

> Ontology does not occur at a level reserved for principles, a level that is withdrawn, speculative, and altogether abstract. Its name means "the thinking of existence." And today the situation of ontology signifies the following: to think existence at the height of this challenge to thinking that is globalness as such (which is designated as "capital," "(de-) Westernization," "technology," "rupture of history," and so forth).
>
> —JEAN-LUC NANCY, *Being Singular Plural*

> The unthinkable is not something we are thinking about at the moment.
>
> —PETER KENYON, chief executive of Manchester United Football Club

In the Fine Underwear of Our Minds

There is a scene in Tom Tykwer's rather pedestrian film *Run Lola Run* (1998)[1] when the two protagonists—Lola (Franka Potente) and her boyfriend, Manni (Moritz Bleibtreu)—lie in bed discussing the random nature of love and existence. Lola asks Manni that age-old question which lovers often pose their partners: "Why me? Of all the people in the world, why did you pick me?" As we know, one of the main functions of romantic narrative is to weave all the strands of coincidence and contingency together in such a way that the lovers feel compelled to believe in the benign intervention of an invisible hand of fate. "It could not have been otherwise," they tell each other. (If it *could* have been otherwise, then this counterfactual eventuality would have erased the first possibility, thereby permanently sealing the other road not taken.)

"Why me," asks Lola, "and not one of those other girls?" Her boyfriend responds as the ancient script demands, reassuring Lola of those unique qualities that ensure her status as the loved one: the individual who stands out amongst other individuals. Imagine, however, that we were to indulge in some retrospective script doctoring, delicately rewriting this scene according to a less-established romantic formula. In such a case, Manni's answer could have been "But Lola, you *are* one of those other girls."[2] Such a subtle shift in perspective may seem trivial, pedantic, and even a little cruel in the context of ego reassurance. Nevertheless, emerging conceptual models of "being-in-the-world" are forcing us to rethink the relationship between self and other, subject and object, individual and community in ways which are anything but trivial. How the subject negotiates the highly fluid character of contemporary society and orients him- or herself within the schizo-semiotic flux of the twenty-first century have become questions of pressing concern to those with an interest in deconstructing the relatively stable (and staple) Freudian-Enlightenment unit of ego-based individuality.

The seeds of this emerging perspective have recently been harvested by the philosophical investigations of community by Giorgio Agamben, Maurice Blanchot, and Jean-Luc Nancy, although they were sown several decades earlier in the sprawling literary fields of Marcel Proust and Robert Musil. As is the way with agricultural metaphors, such seeds can be traced back to the earliest articulations of art and philosophy.

In *Remembrance of Things Past*, Proust's narrator, Marcel, travels to the seaside town of Balbec, where he is confronted by a gestalt gaggle of young women walking along the sand:

> Although each was of a type absolutely different from the others, they all had beauty; but to tell the truth I had seen them for so short a time, and without venturing to look hard at them, that I had not yet individualised any of them . . . and when (according to the order in which the group met the eye, marvellous because the most different aspects were combined in it, but confused as a piece of music in which I was unable to isolate and identify at the moment of their passage the successive phrases, no sooner distinguished than forgotten) I saw a pallid oval, black eyes, green eyes emerge, I did not know if these were the same that had already charmed me a moment ago, I could not relate them to any one girl whom I had set apart from the rest and identified. And this want, in my vision, of the demarcations which I should presently establish between them permeated the group with a sort of shimmering harmony, the continuous transmutation of a fluid, collective and mobile beauty. (1989, 847–48)

This "pale madrepore" constitutes an "invisible but harmonious bond, like a single warm shadow, a single atmosphere, making of them a whole as homogenous in its parts as it was different from the crowd through which their procession gradually wound" (851).

When Marcel is later shown an old photograph of these childhood friends, he notes that "those children, still mere babies, had been at that elementary stage in their development when personality has not yet stamped its seal on each face. Like those primitive organisms in which the individual barely exists by itself, is constituted by the polypary rather than by each of the polyps that compose it, they were still pressed one against another" (882). At this stage in the girls' development, something as evanescent as a giggle is enough to dissolve the ontological boundaries between them, "obliterating, merging those imprecise and grinning faces in the congealment of a single cluster, scintillating and tremulous."

When merely children, these girls are incipient creatures, awaiting the "stamp of personality." Nevertheless, even some years later and in the flush of early womanhood, the sheer plurality of their presence overwhelms Marcel. While in Balbec "each of their physiognomies was now mistress of itself" (882), it takes a while before the narrator's libidinal and phenomenological excitement calms down enough to sift each girl out from the other. In contrast to Levinas's "epiphany of the face"—that unavowable confrontation with the other's incommensurability—Marcel experiences a confusion of features, a multiplication of qualities which suspends any kind of ethical relationship.

It is within the context of this sensory overload that we can approach the question of communicative identity. As Walter Benjamin has shown, the public spaces of the late nineteenth and early twentieth centuries confronted the subject with an unprecedented amount of information to process and interpret. The boundaries of the self soon became threatened by the semiotic missiles hurled by advertising men, architects, and city planners, designed to penetrate the consumer-citizen. While Balbec is not exactly Times Square, it too is prey to the confusional order of comingling classes and types. This particularly modern form of discordant order follows the techno-logic underpinning contemporary quotidian life and the fleeting encounters that it promises. Marcel meditates on the effect produced by women glimpsed from train carriages, especially how the charm of a girl can often be measured in direct inverse ratio to the amount of time she is in view. He reflects on the "evanescence of persons who are not known to us" and laments the "numberless strangers whom, even at Balbec, the carriage bowling away from them at full speed had forced me for ever to abandon" (853). In an age when the tempo of life is gaining pace in all quarters and the earth itself seems to be spinning faster on its axis, technology, and technological metaphors, seem best equipped to capture the acceleration of perception. Accordingly, Balbec beachgoers are forced by that dazzling gang of young women to move "as though from the path of a machine which had been set going by itself and which could not be expected to avoid pedestrians" (848).

Robert Musil's man without qualities, Ulrich, is also finely attuned to the new kinetic geometries of city life. We first meet him standing behind a window, "ticking off on his stopwatch the passing cars, trucks, trolleys, and pedestrians, whose faces were washed out by the distance, timing everything whirling past that he could catch in the net of his eye. He was gauging their speeds, their angles, all the living forces of mass hurtling past that drew the eye to follow them like lightning, holding on, letting go, forcing the attention for a split second to resist, to snap, to leap in pursuit of the next item" (1996, 6).

We are therefore introduced to Ulrich as a kind of narrative Archimedean point, the only still mechanism in the urban machine. But this is something of a red herring and should not lead us into the traditional assumption that Ulrich represents a stable subjectivity, someone who can successfully negotiate the bustling activity of the street and absorb the shocks of city life by harmlessly incorporating them into his person. Musil's protagonist represents a distinctly modern (or perhaps postmodern) attitude to his environment. As a consequence, he is not the type to use each info-missile as a pertinent reminder of the flesh, which acts as the border between himself and others, between Ulrich and not-Ulrich.[3] Moreover, his identity adapts to the semiotic swirl by becoming indistinct and indefinite. "It is not difficult to describe the basic traits of this thirty-two-year-old man Ulrich, even though all he knows about himself is that he is as close to as he is far from all qualities, and that they are all, whether or not he has made them his own, in a curious fashion indifferent to him" (159).

Indeed, when he gets into some trouble with the police and comes face to face with the machinery of the state, Ulrich perceives his interrogation in similarly technocratic terms:

> He felt as though he had been sucked into a machine that was dismembering him into impersonal, general components before the question of his guilt or innocence came up at all. . . . His face counted only as an aggregate of officially describable features—it seemed to him that he had never before pondered the fact that his eyes were gray, one of the four officially recognized kinds of eyes, one pair among millions. . . . He could, even at such a moment as this, himself appreciate this statistical demystification of his person and feel inspired by the quantitative and descriptive procedures applied to him by the police apparatus as if it were a love lyric invented by Satan. The most amazing thing about it was that the police could not only dismantle a man so that nothing was left of him, they could also put him together again, recognizably and unmistakably, out of the same worthless components. (168–69)

Ulrich is thus something of an enigma to people like Paul Arnheim, the rich industrialist, intellectual, and philanthropist who embodies the man *with* qualities. All of Arnheim's decisions are based on his belief that it is "far from impossible that a great, superrational collectivity was coming to birth and that, abandoning an outworn individualism, we were on our way back, with all the superiority and ingenuity of the white race, to a Paradise Reformed, bringing a modern program, a rich variety of choices, to the rural backwardness of the Garden of Eden" (444). (This is something of an irony, considering that Arnheim is a constant focus of the media's cult of personality.)

Such a vision, however, rests not on a dissolution of identity but on a humanistic faith in communion as a sacred form of unification. For Arnheim the twentieth century leads inevitably to a secular fusion of values and ideals, as embodied by the ideological ambitions of the European Union. Ulrich, on the other hand, anticipates the Deleuzian model of schizoanalysis, the complex fragmentation of subjectivity in the face of the Spectacle. Thus, we

> should remember that having a split personality has long since ceased to be a trick reserved for lunatics; at the present-day tempo, our capacity for political insight, for writing a piece for the newspapers, for faith in the new movements in art and literature, and for countless other things, depends wholly on a knack for being, at times, convinced against our own convictions, splitting off a part of our mind and stretching it to form a brand-new whole-hearted conviction. (424)

As Musil reminds us, in the age of rampant capitalism, "what people are" evidently keeps changing as rapidly as "what people are wearing" and that "no one, not even those in the fashion business, knows the real secret of who 'these people' are" (494). Those economically privileged people who, in the "fine underwear of their minds," once knew who they were and where they belonged, as surely as if it were monogrammed on their souls, are already in jeopardy in Musil's prewar Vienna.

Significantly, Musil's friend and fellow writer Hermann Broch takes up this same theme in his own major work, *The Sleepwalkers.* Broch's initial protagonist, Joachim von Pasenow, is an army lieutenant and, as such, a member of that "cult of the uniform" which transforms the wearer "into a property of his uniform" (2000, 14). The uniform is thus a hard casing, as much psychological as material, which is used to "arrest the confusion and flux of life, just as it conceals whatever in the human body is soft and flowing, covering up the soldier's underclothes and skin" (15). Indeed, when in uniform, Joachim "begins to forget his own undergarments," and along with them the uncertainties of modernity. Fighting a constant border war on "the frontier between his self and his uniform," Joachim is tortured by the shame which he associates with life outside the military for the simple reason that "everything civilian" is "a matter of underclothing" (16).

And so, between Ulrich and Joachim, Musil and Broch, underwear and uniforms, we see the emergence of the postdeterministic character of a modern subjectivity without qualities. The "dehumanizing" agenda of certain globalist tendencies, where the qualities of one person can seemingly be recombined with another (even beyond the laws of physics, physiology, and cosmetic surgery), thereby becomes the site of an antiessentialist struggle over the very notion of identity. Writing between Nietzsche and Foucault, these literary prophets helped trace the various faces of Western individuality in the very sand which awaits the erasing wave of history.

Whateverbeing

The Irish tourist bureau has a saying: "A stranger is just a friend you haven't met yet." From a different perspective, however, a friend is just a stranger who happened to cross your path. (This may also explain why strangers are very rarely strange, at least within one's own circles and experience. Indeed, it may make more sense to call them "strangely familiars.")[4] Italian philosopher Giorgio Agamben has coined the term "whatever being" (*qualunque*) for the emerging postsovereign subject. This whatever being is the mode for the "coming community," a rather abstract theoretical blueprint for a way of imagining life between the metaphysical landmarks of becoming, being, and belonging. The crucial point to remember in the context of this discussion is that the coming community is based on an "*inessential* commonality" (1993a, 18). According to Agamben, then, "coming being" represents pure possibility: Humankind "has to *exist* as *potentiality*" (1, 44).

On a slightly more pragmatic level, this entails a fundamental revision of what it means to be a person: to declare that uncoded existence precedes the modern circumscriptions of citizenship, family, religion, ethnicity, and other blood-soaked calls to an essential identity. According to such a perspective, even the United Nations' alleged mandate of defending "human rights" colludes with the tyranny of essentialist discourses, smuggling all sorts of assumptions about human nature across the disputed borders of the planet.

One of the motivations behind my decision to write this book began with a seemingly simple question: What does Agamben mean when he talks about "whatever being"? As is usually the case with seemingly simple questions, they lead to other questions, which themselves spiral out into a network of interrelated and ultimately more complex questions. How does this notion of whatever being relate to Agamben's concept of the coming community? What would an "inessential commonality" (the basis of such a community) actually look like? Is the "automatism of love" (Zizek) a stumbling-block to thinking the coming community, or is there a way to reinflect the lover's discourse beyond notions of subjective essence? And how do contemporary forces or phenomena—such as the media, globalization, and the information revolution—encourage or discourage an emergent conception of community outside the exhausted coordinates of humanism, the humanities, nihilism, postmodernism, and psychoanalysis (to name only a few)?

It is this cluster of questions which connect the different chapters of this book as they pursue the conceptual antimatter of Agamben's critical project, which—sometimes maddeningly—"consists not in discovering its object but in assuring the conditions of its inaccessibility" (1993b, xvi).[5] Whether it be via cinema, software, literature, photography, or philosophy, the following case studies seek an encounter with emergent postmillennial conceptions of community, neither beginning nor ending with the subject but flowing through the increasingly obsolete circuits of ego-centered individuality. This enables us in turn to develop the the mental muscles necessary for thinking whateverbeing. Humanist metaphysics insists we are all unique and yet share an essential, universal humanity. A caricature of the coming community suggests a different approach: None of us are unique, yet there is no point which we all share.

The task at hand, therefore, is twofold: to trace the emergent conceptions and amplifications of an anticipated coming community and to map the textual reworkings of "love" on the level of an a-subjective subject. Indeed, Agamben tells us:

> Love is never directed toward this or that property of the loved one (being blond, being small, being tender, being lame), but neither does it neglect the properties in favor of an insipid generality (universal love): The lover wants the loved one *with all of its predicates*, its being such as it is. The lover desires the *as* only insofar as it is *such*—this is the lover's particular fetishism. Thus, whatever singularity (the Lovable) is never the intelligence of some thing, of this or that quality or essence, but only the intelligence of an intelligibility. The movement Plato describes as erotic anamnesis is the movement that transports the object not toward another thing or another place, but toward its own taking-place—toward the Idea. (1993a, 2)

If we return to Lola and her nocturnal conversation with Manni, the key question then becomes: How does love figure in the coming community? If we are all men and women without qualities, then according to what criteria do we select one partner over another?

(It is no doubt prudent to clarify that I do not ask this question in order to find a psychological or biological explanation for object choices. While I think it is misguided to underestimate the power of pheromones, or indeed the Oedipus complex, in the great arena of human seduction, I am more interested in the "color of love" when it is dropped into a solution comprising equal parts literary, visual, and media culture.)

Fortunately, Thomas Carl Wall's book *Radical Passivity* (1999) has provided us with a useful metaphor which we can apply to the mysteries of Eros. Wall's example of whateverbeing (hereafter fused into one word) is the character actor, that cinematic stereotype who remains "so unknown to us not because they hide an essence, but because they are completely exposed"

(133).[6] Unlike the movie star, whose particularly magnetic qualities draw us toward the silver screen, the character actor is "the pure 'taking place' of those qualities: an actor = x, sort of" (138). As Wall reminds us, nobody goes to the movies to see a Thelma Ritter movie. The character actor is therefore an example of whateverbeing, an enactment of existence without qualities, or at least qualities so interchangeable and obvious that they erase all identity. This is not to be confused with an essence, or existential common denominator, but rather the sheer generic potentiality of being. The character actor is "nothing other than its qualities but such that these qualities cling to no reality, no identity, and refer only to themselves" (134).

Let us consider another recent film which thematizes the blurring of boundaries between self and other, Spike Jonze's *Being John Malkovich* (1999). In this story a bedraggled puppeteer, Craig Schwartz (John Cusack), discovers a portal into John Malkovich's brain, allowing anyone access to another subjectivity for at least 15 minutes (which may or may not have a Warholian subtext). The puppeteer's wife, Lotte (Cameron Diaz), soon becomes addicted to this experience, claiming that she only feels truly "herself" when she is inside John Malkovich's head. "But you weren't you; you were John Malkovich," the puppeteer reminds her. Yet, she remains unimpressed by such pedantic distinctions.

While the notion of mind-hopping and body-jumping is also the basis for countless identity-swap films, *Being John Malkovich* presents the self as something far more fluid, vulnerable, and communicative than the average Hollywood depiction. Wall tells us:

> Every statue, every cadaver, every puppet, toy, or artifact—indeed every thing and every person who falls, if only for a moment, outside utility—returns to an inconceivable image void of either subject or object. . . . They invert the "motion of the world" and "return us to ourselves" [Blanchot], but to ourselves insofar as there is no one to return to, no society of identities in which we can recognize ourselves. (110)

"What remains," continues Wall, "is pure *-jection* (or throwness)" (110).

From this Heideggerean standpoint, the puppeteer fuses with the puppet as he is *thrown* down the portal into the ontic space of another human vessel. The fear of rejection therefore acquires an altogether different meaning in this film: the literal fear of being repetitively thrown back into the agonizing loneliness of the self, rather than the dialectical dance of the other. Hence the repetition throughout the film of the puppeteer's tortured "dance of despair" (dejection), in which the puppet smashes a mirror in a bid to erase the solipsistic reflection of the self.

Deleuze and Guattari anticipate such a *mise-en-abyme* when they write:

> Puppet strings, as a rhizome or multiplicity, are tied not to the supposed will of an artist or puppeteer but to a multiplicity of nerve fibers, which form another puppet in other dimensions connected to the first: "Call the strings or rods that move the puppet the weave. It might be objected that its multiplicity resides in the person of the actor, who projects it into the text. Granted; but the actor's nerve fibers in turn form a weave. And they fall through the grey matter, the grid, into the undifferentiated. . . . The interplay approximates the pure activity of weavers attributed in myth to the Fates or Norns." (1999, 8).[7]

Indeed, *Being John Malkovich* zooms in on the fine lines separating self-directed love from other-directed love, exposing the artificial nature of such a binary.[8] Take, for instance, the scene where John Malkovich slides down his own portal, emerging into a dream restaurant where every person wears his face. This nightmarishly narcissistic realm tortures Malkovich, for though he is surrounded by other people, they are all "Malkovich." Moreover, this explosive solipsism (which we could call "Aphex Twin syndrome," after the avant-garde musician whose video clips have perfected this abject form of *me-jecting*) both destabilizes and reinforces the eternal conflict between attraction and repulsion which informs sexual attraction from infancy to senility.[9]

Once again, such a conflict prompts many questions: Am I looking for a piece of myself in the other, or the other in myself? Is the romantic notion of the perfect couple merely "two views of the same person" (DeLillo 1985)? Or is the yin-and-yang simplicity of modern monogamy, however fleeting, an attempt to control the sheer infinite potentiality of human combinations?

Go-Sees

There is a photography book currently on the market entitled *Juergen Teller Go-Sees*, a collection of nearly five hundred shots of budding fashion models, captured by the *über*-hip Teller (1999). In contrast to most glossy coffee-table books related to the glamor industry, these are candid shots of young women who come knocking at Teller's door in the hope of becoming the next Kate Moss, photographed previous to their receiving any styling treatment as models.

The blurb is an interesting case of cultural-theory-as-advertorial:

> These are photographs of arrival and departure, portraits of brief but loaded encounters that articulate underlying continuities and discontinuities. The changing texture of light registers the passing seasons, while the recurrent backdrop hints at an overarching narrative. The models become characters in a fiction that the viewer is invited to imagine. In the course of the book the doorway, as a charged liminal space, emerges as a site of meeting and contest, at once enticing and forbidding: a point of rupture and a contact zone. . . . The doorway, as border, provides a visual analogy for the ambivalent politics of inclusion and exclusion. The photographs demonstrate how, in the fleeting interstitial space of the go-see, the body can momentarily elude the constraining stereotypes of a world in which stardom is the longed-for exception. (n.p.)

In my experience, the most common viewer reaction to this collection of photographs is an expression of being overwhelmed. Face follows face, until it gets so each model simultaneously cancels and converges into a palimpsest of waifish features without qualities. Like Proust's Marcel, we are confronted by "a sort of shimmering harmony, the continuous transmutation of a fluid, collective and mobile beauty," or better yet, of a piece of music in which we are "unable to isolate and identify at the moment of their passage the successive phrases, no sooner distinguished than forgotten."

In *Go-Sees* we have a perfect example of Wall's notion of the character actor and its relation to whateverbeing. These models represent "constraining stereotypes of a world in which stardom is the longed-for exception," an exception which necessarily negates its own desires. The model seeks fame in order to enhance the egocentric delights of wealth and attention. And yet this attention isolates the model further and further, so that she is a cutout in sharp relief to the rest of the population, placed in a spotlight or upon a pedestal. The celebrity's wish for anonymity is the inevitable consequence of fame, since love cannot adhere to such a strong social substance as stardom and to the fetishization of "personalities."

The burgeoning popularity of candid street-portrait photography is a testament to the inchoate cultural acknowledgment of whateverbeing. Japan in particular has produced magazines—such as *Fruits*, *Kerouac*, and *Street*—which do not so much celebrate the idiosyncratic plumage of the individual as champion the malleable fundament of human being. As we shed the shell of Enlightenment individualism, these modern-day flip books reveal more than one might imagine about the dialogue between different pop cultures. From a certain angle these photographs tentatively capture a coming community based on sheer commensurability, the spectacular circulation of qualities which today's city dweller prefers to wear than absorb. (Recall Musil: "What people are" evidently keeps changing as rapidly as "what people are wearing.")

Yet, *Go-Sees* also reminds us of the gendered aspect of my reading of whateverbeing. To be overwhelmed—as we are, flipping glossy page after glossy page or watching the girls on Balbec's beach—is itself a Romantic trope par excellence, bequeathed to us by such skilled literary programmers as Flaubert and his search for the perfect gynaeceum. The vision of being surrounded by a sea of sexual creatures, figured in the feminine, is an explicitly erotic mobilization of the sublime, an oceanic trope which, as we have seen, now has its urban equivalents. While it is important to remember the constitutive character of the masculinist gaze in the texts mentioned above, it is not difficult to find examples of the "female" deployment of this particular motif (for instance, countless television commercials for luxury-end chocolates).

The camera is thus one of the latest in a long line of technologies (including language itself) which has the capacity to cut the other into pieces, specifically for the thrill of rearranging personal qualities and putting them back together differently in a collage. In the late-twentieth-century obsession with recombinant identities—beginning with Proust and Musil, among many others, and extending to Teller, Tykwer, and Jonze—we see an emerging renegotiation of embodied identity in an unprecedented technophilic and -phobic age.

While one of the intentions of this book is to find "examples" of whateverbeing, it does so in the knowledge that Agamben has warned against such an exercise, since this concept necessarily resists its own representation. For its founding architect, the coming community is neither prophetic nor prescriptive, but works against such deterministic models (much like Derrida's "messianism without a messiah"). In fact, the coming community is not a model at all but an antimodel. With this in mind, the chapters which follow do not simply offer concrete incarnations of whateverbeing but simultaneously dwell in those potential and emergent spaces where the elements of its figuration take precedence.

Indeed, whateverbeing may be like those messages beamed by lighting devices which rely on creating negative afterimages in the eyes; we may have no idea what such a message is trying to tell us until we look away. Those glimpses of whateverbeing that we manage to catch in the following pages whisper to us of *les arrivants*, the ones who perpetually approach but will never arrive. But expectation itself—provided it partakes in the informed politics of exhaustion, as opposed to the historical reflex of utopianism—can fruitfully deploy itself within real-world conditions. Whether it be through Lolita, Cordelia, Ulrich, Hong Kong drifters, or Los Angeles pornographers, the undoing of violent essentialism becomes a powerful figure through which to comprehend the process of becoming itself.

Chapter 1 begins by outlining key terms and introducing the conceptual instruments which will be used to forge *eros*, *technics*, and *communitas* into a passionately turbulent (re)union. This discussion encompasses Plato, Roland Barthes, Bernard Stiegler, and Niklas Luhmann, among others making cameo appearances, in order to highlight the ways in which "the lover's discourse" functions across different epochs, despite several critical flaws.

Chapter 2 continues to pick at these immanent contradictions in the romantic code by focusing on Vladimir Nabokov's canonical text, *Lolita*. The character of both Humbert Humbert and the nymphet herself are explored via the matrices of mnemotechnics, the passionate compulsion to write, record, and otherwise "salvage" the beloved from the ravages of time. Particular attention is paid to the different figurations of Lolita by Humbert, so that the novel itself is read as an instance of science fiction or, more specifically, an erudite essay on the practical possibilities of time travel.

Chapter 3 turns its attention to the corporate business park of J. G. Ballard's *Super-Cannes* in order to identify the contemporary macro conditions of whateverbeing. In this novel a new caste of elites inflicts a disciplined and externally applied form of psychopathology against the local immigrants, as a salve against a life dominated by work rather than emotional attachments. Indeed, Ballard's novel is a revealing treatment of certain dangers inherent in the politically charged nonspaces of today's vocational environments.

Chapter 4 updates Deleuze's concept of "faciality" to a more networked notion of "interfaciality," examining the way the face has been traditionally considered as a marker, or mask, of a singular soul. This discussion incorporates new technologies such as digital avatars, recent research on the biological mechanics of perception, films such as *Face/Off* and *Seconds*, William Gibson's novel *Idoru*, as well as Levinas's approach to "the epiphany of the face." And it does so in order to reinject an awareness of ethics, albeit one untethered from the traditional criteria of sovereign subjectivity and more aligned with open membership in the coming community.

Chapter 5 argues that "mainstream" pornography enacts the fundamental tension of whateverbeing—that is to say, the *structural interchangeability* of the "beloved" with the simultaneous *radical irreplaceability* of the same. Incorporating Zizek's notion of "inter-passivity" as well as Agamben's thoughts on the sexualized, post-auratic body, this section finds evidence for the (perhaps surprising) claim that a "saving power" grows within the "danger" of pornography, since it explicitly frames and thematizes the "inessential commonality" of desire.

Chapter 6 gathers together the themes of belonging, identity, and desire and reads them through the rigorous attempts to rethink community by Jean-Luc Nancy, Martin Heidegger, and Giorgio Agamben, particularly those thoughts dedicated to the role of the lover's discourse in this new ethical assemblage. Such a synthetic balancing act is supplemented by Gilles Deleuze's particularly intriguing notion of *essence* from his book *Proust and Signs*, which I use to counter the increasingly ritualistic denunciation of "essentialism" in the academy.

Chapter 7 builds on this new foundation in order to consolidate our understanding of the "immanent isomorphism" of love, technology, and community. As such, Irigaray's more recent work on the ontology of love is used as a foil to my own argument, since she exhibits a neo-Luddite approach to technology, as if it could simply be quarantined from such an allegedly pure and innocent sphere as the human heart. The work of such contemporary critics as Steven Shaviro and Mark Poster is then enlisted to provide a more nuanced account of how the conceptual status of the subject is being recoded by actual new-media practices and formations.

Chapter 8 reinforces these ideas under the assumption that humans are, at the most fundamental level, "culturally accented" (Naficy), and thus we should be wary of metaphysical generalizing. The films of Wong Kar-wai as well as the writings of Haruki Murakami are introduced in order to tease out some of the contemporary poetics of global techtonics. These "fallen angels," I argue, mark the current state of our subliminal thinking on whateverbeing, which unfolds within the context of cultural diaspora and disorientation.

The conclusion, for its part, ties all these threads together, re-presenting the basic premise that current conceptions of love and community are hampered by our neglect of a third term, namely, technology. *For it is emphatically not a case of technology rampaging out of control, crushing human nature in its machinic path, but rather a situation wherein the human is constantly recreating itself, from the very beginning, as a technical animal.* The erotic encounter is inevitably enframed by technologies, from architecture to etiquette. Even if we happen to fall head over heels for someone in the forest or on a beach, naked, there is the *über-techne* of language to take into account. And where there are two (as Friday's footprint told Robinson Crusoe), there is a virtual community. In other words, there is no human outside technology or, for that matter, outside the legacy of love and community. We are cut from this same cloth, and the first step toward reverse-engineering a flexible future is to recognize the way these terms are each implicated in the others.

ONE

Love and Other Technologies

Love is the sign that one is changing discourses.

—JACQUES LACAN, *On Feminine Sexuality*

Let us imagine the first handshake. This gesture, ubiquitous in the West and therefore increasingly common throughout the world, extends back into the mythological mists of commerce and community. The handshake signals a greeting, an introduction, a contract made, and an understanding achieved. In a quotidian version of Michelangelo's finger-touch between God and Adam on the ceiling of the Sistine Chapel, one being confronts another and makes *contact*. A border is crossed (the gulf separating two persons) and is simultaneously affirmed (since the two pressed palms continue to mark the point at which one person is distinguished from the other). The handshake therefore serves as a reverse mirror stage in which two subjects recognize themselves as distinct and attempt to bridge the gap.

The same could be said for the kiss, a more amorous version of the handshake, in terms of the will to extend the self toward the other. Love "itself" traces the movement—or perhaps *is* the movement—of this will to connect, to extend and incorporate. It is a language of the social fundament which enables all other languages through this perpetual will to communicate. Language is a technique of interaction and thus qualifies as a form of technics.[1] As such, love *is* a technology. But to understand this point—that *love, technology,* and *community* are in fact terms which designate the same phenomenon or process—we must be more forthcoming with our own conceptual tools.[2]

I base this caricatured handshake on Bernard Stiegler's history of the human in his book *Technics and Time* (1998). Stiegler's study presents an argument for an invention of the category "human" through a rereading of Rousseau, Simondon, Leroi-Gourhan, Heidegger, and the myth of Prometheus and Epimetheus. This study begins with a portrait of Rousseau's Original Man, who is completely autonomous and within himself and thus has no means of a social mode of interaction with others.[3] We imagine this Original Man sitting under a tree, perhaps eating a banana, untroubled by mental activity above the level of vague physiological registerings of pleasure and pain. This man "carries himself, as it were, perpetually whole and entire about him" (Rousseau), needing only food and warmth and other pre-linguistic behaviors unsullied by (self)consciousness. "Everything is inside," Stiegler tells us, "the origin is the inside" (116).

However, there comes the evolutionary point where originary prehumans want to communicate with each other and follow the bioskeletal logic of their feet, hands, voice boxes, opposable thumbs, and facial features. Hence the emergence of various prostheses: speech, grasping, reaching. The hand and the tongue become tools; in fact, they were not anything we could designate as "hand" or "tongue" before this becoming. And so begins Original Man's fall into exteriorization, a fall which marks the point of emergence of the human from the nontechnical interior world of animals. Indeed: "This thematic of exteriorization is central to . . . the process of humanization" (Stiegler, 1998, 116). Through this rewriting of Rousseau's fable of human origin, Stiegler makes some extremely useful points in thinking through the notion of technology as a movement with fundamental zoological, temporal, and ontological implications. For if "speech is already a prosthesis," then "[a]ny exit outside of oneself is a denaturalization" (116). The human is now at one remove from the natural state against which it distinguishes itself. "Hominization . . . means that the appearance of the human is the appearance of the technical" (141).

It is crucial to understand this concept of exteriorization as not simply a move from inside to outside but a trajectory which creates *this very distinction* through the movement itself. This is an example of Simondon's "transductive relation": "a relation which constitutes its terms, in which one term cannot precede the other because they exist only in relation" (in Derrida 2002, 161). In other words, there was no "inside" before the moment of exteriorization which created an "outside": "Interior and exterior are consequently constituted in a movement that invents both one and the other: a moment in which they invent each other respectively, as if there were a technological maieutic of what is called humanity" (Stiegler, 1998, 142). To put it another way, interior and exterior form "an originary complex in which the two terms, far from being opposed, compose with one another (and by the same token are posed, in a single stroke, in a single movement)" (152).

According to this schema, the human *is* technics, and we cannot romanticize a once-upon-a-time where humanity was unsullied by its artificial creations. There was no Edenic moment before the Fall, or at least not one which could be considered a part of human "history." For if the hand itself is a tool and speech a technology, then the human is always already a technical animal. Thus, the "fall" into our social state is not simply a fall into technology but more specifically a fall into time itself. Original Man was unaware of death as a part of his being, since without the technology of prolepsis (language, grammar, culture), he could not anticipate. So, the human emerges simultaneously with a comprehension of the self as one among many, a being among beings, each one conscious of death.

Stiegler goes on to inform us that "[s]ince original man has neither imagination nor future, nor for all that a memory or a past, he is almost without love, almost without desire" (131). What we now call passion is unknown to Original Man, since, to quote Rousseau, "every woman equally answers his purpose" (in Stiegler, 131). Love can therefore emerge only within the moment and movement of exteriorization, which, as we have seen, is itself technical. Love is thus a technology of community, an acknowledgment and recognition of alterity and affiliation. Alternatively, technology is a love of community, Freud's mythical Eros engineering and building enduring social relationships through time. Or further still, community is the love of technology and the technology of love—a recombinatory ménage à trois, which itself reflects the crucial third term in thinking through the discourse of belonging.

But why love? Is this not assuming a lot from a simple handshake? After all, deals sealed in this way often lead to double crossing, and humans seem as quick to clobber each other over the head as to perform a dawn serenade beneath a beloved's window (as already suggested by Kubrick's violent, bone-wielding man-ape). For the moment I do indeed use *love* in its most loose form, as the force of Eros and agapē, or the ideal of a general goodwill which allows the fostering of communities. Love, in this sense, is a philosophical and social principle tied to ethics, allowing a sense of belonging which is fundamental to the community and applicable to the abstract category of humankind. Another word for this kind of love could be *empathy* or *sympathy*, which overlap and speak of a kind of invisible medium which allows one person to feel for or put him- or herself in the place of another person.[4]

Empathy is a secretion of the process of exteriorization described above, since it is impossible to have one without the other. Like the great coral reefs, which are formed of the exuded exoskeleton of the multitude, empathy is the social glue which allows us to "stick together." In a complex feedback loop, the countless pathways and networks of social commerce follow the empathic circuits which condone the handshakes that encourage the economy in the first place.

The question then arises: Are social subjects "in" empathy the same way they are "in" time, or "in" culture? We shall answer hastily in the negative and hope to justify the answer throughout the course of the following chapters, since we are aware of the dangers of thinking in such terms. Just as Benjamin warns against the simplifications of "empty homogenous time" (and McLuhan collapses the medium with the message), we must avoid the temptation to see subjectivity as somehow prior or external to the media of ethics, language, time, technology, and culture.

But again, what of love? Can we consider ourselves living within "empty homogenous love"? Clearly, falling "in" love is linked to the mythical fall "into" time, but the details remain sketchy at this point. If the human is indistinguishable from technics, then it is also indistinguishable from time. The task is thus to rethink the link between subjectivity and love, with special attention to the ways in which the former has been figured, prefigured, and configured by technology, temporality, and cultural transmission. Only then can we recode the lover's discourse into something unhampered by the metaphysical exigencies of "backwards compatibility" (i.e., the subtle violence and hypocritical myopia of the romantico-humanist mode).

A Thousand Platos

> And if we are not obedient to the gods, there is a danger that we shall be split up again and go about in basso-relievo, like the figures having only half a nose which are sculptured on columns, and that we shall be like tallies.
>
> —ARISTOPHANES, in Plato, *The Symposium*

Few terms are as sticky, slippery, or treacherous as *love*. We will not, therefore, try to define "love itself." Should we happen to make a gesture toward definition in passing, such a gesture will be acutely aware of its status as only one among an infinite set of possible meanings. Rather, our focus is to unpack the notion of love as a *discourse*. In other words, the question is not "What is love?" but rather "What is loving?" However we choose to label this phenomenon—romance, love, lust, affection, passion, intimacy, eroticism, longing, dependence, desire, sexuality, obsession—we are dealing with a *discursive* constellation. It is constituted through discourse, embedded in the different cultural and historical genealogies of that-which-is-called-love. Hence, one of our key questions arises: How can we best think the link between sexuality and textuality?

Another name for the lover's discourse, as defined by Roland Barthes, is "the image-repertoire." The image repertoire precedes us, a fact which led La Rochefoucauld to say, "People would never fall in love, if they had not been told about it" (in Polhemis 1990, 121). The image-repertoire is the discursive legacy of love stories which we have imbibed and internalized since we were children, often through literature but also through television, movies, gossip, fairy tales, magazines, and so on. The image-repertoire can be considered the world data bank of images of love, with which we construct our own imaginative mise-en-scènes.

To tweak the metaphor a little, the image-repertoire is a *code*. Like a computer code, it is designed in order to execute a program, here inscribed into society, in order to keep things running. But like a secret code, we must learn to *de*-code it. "Each of us can fill in this code according to his own history," writes Barthes, "rich or poor, the figure must be there, the site (the compartment) must be reserved for it" (5). We will come back to this notion of the codification of love throughout the book, but for now I would simply introduce the notion of a set of operating instructions for the way we behave in an intimate manner.

One of the first programmers of this code is Plato, especially in his *Symposium,* although many contradictory opinions on love are voiced in this text. According to Phaedrus, it is the oldest of forces: "First Chaos came, and then broad-bosomed Earth, The everlasting seat of all that is, And Love. In other words, after Chaos, the Earth and Love, these two, came into being" (705). In contrast, Agathon believes Love to be the youngest of the gods, "and youthful ever." In response, the physician Eryximachus views the debate through his own vocation:

> Medicine may be regarded generally as the knowledge of the loves and desires of the body, and how to fill or empty them; and the good physician is he who is able to separate fair love from foul, or to convert one into the other; and if he is a skillful practitioner, he knows how to eradicate and how to implant love, whichever is required, and he can reconcile the most hostile elements in the constitution, and make them friends. (715)

Medicine here—above all a technology or technique—requires a profound understanding of love and in fact is inseparable from the regulation and manipulation of love. In turn, this treatment is for the benefit of not only the individual body but also the social body, so that we see once again the inextricable coiling of love, technology, and community.

Following on from these words by Eryximachus, Aristophanes (who had to delay his thoughts on love due to a kind of hysterical hiccoughing reaction to the discussion itself) regales the table with the fable of original androgynous existence, a tale which would subliminally saturate all succeeding discussions of love in the West. According to Aristophanes:

> In the first place, the sexes were originally three in number, not two as they are now; there was man, woman, and the union of the two, having a name corresponding to this double nature; this once had a real existence, but is now lost, and the name only is preserved as a term of reproach. In the second place, the primeval man was round and had four hands and four feet, back and sides forming a circle, one head with two faces, looking opposite ways, set on a round neck

> and precisely alike; also four ears, two privy members, and the remainder to correspond. When he had a mind he could walk as men now do, and he could also roll over and over at a great rate, leaning on his four hands and four feet, eight in all, like tumblers going over and over with their legs in the air; this was when he wanted to run fast. (719)

Zeus is appalled by the noisy insolence of these hermaphrodites and decides to teach them a lesson by cutting them in half:

> After the division the two parts of man, each desiring his other half, came together, and threw their arms about one another eager to grow into one, and would have perished from hunger without ever making an effort, because they did not like to do anything apart. . . . Each of us when separated is but the indenture of a man, having one side only like a flat fish, and he is always looking for his other half. (720–21)

In a key moment the mysterious mechanics of love are revealed via the implementation of technology (as figured by Hephaestus, the crippled god of smiths and metallurgy):

> Suppose Hephaestus, *with his instruments*, to come to the pair who are lying side by side and to say to them, "What do you people want of one another?" they would be unable to explain. And suppose further, that when he saw their perplexity he said: "Do you desire to be wholly one; always day and night to be in one another's company? For if this is what you desire, I am ready to melt you into one and let you grow together, so that being two you shall become one, and while you live a common life as if you were a single man, and after your death in the world below still be one departed soul instead of two—I ask whether this is what you lovingly desire, and whether you are satisfied to attain this?"—there is not a man among them when he heard this who would deny or who would not acknowledge that this meeting and melting into one another's arms, this becoming one instead of two, was the very expression of his ancient need. And the reason is that human nature was originally one and we were a whole, and the desire and pursuit of the whole is called love. (722, my emphasis)

Through an ill-defined process linking alchemy with animal husbandry and medicine, Hephaestus's instruments allow us to recognize the persistent discourse of love as the desire for connection, as a pursuit of wholeness, "becoming one instead of two."

Interestingly, it was Hephaestus who chained Prometheus to the rock on Mount Caucasus, once again following the instructions of Zeus. He did not, however, relish this task, admitting to Prometheus in Aeschylus's play *Prometheus Bound*, "With heart as sore as yours I now shall fasten you in bands of bronze immovable to this desolate peak. . . . Your kindness to the human race has earned you this." Further, Hephaestus states, "I hate my craft, I hate the skill of my own hands."

> *Strength:* Why do you hate it? Take the simple view: your craft is not to blame for what must be inflicted now.
> *Hephaestus:* True—yet I wish some other had been given my skill. (1961, 21–22)

But as the discussion concerning Stiegler above shows, the smith's craft is indeed (at the very least partly) to blame "for these present troubles." In order to atone for the forgetfulness of his brother, Prometheus gives humanity the technology of tamed fire. Aristophanes explicitly recognizes the constellation between love, technics, and a peculiarly mortal consciousness.[5] (We should also note in passing that Deleuze and Guattari cite metal as an example of the Body without Organs, and the metallurgist as an essential element of the machinic phylum. Indeed, they go on to claim that smiths are intrinsically ambulant and itinerant, perhaps even forging European space itself via the manufacture of tools, weapons, and the lines of movement that such materials both produce and follow [1999, 410–15]).

As the god of smiths and metallurgy, Hephaestus was responsible for all sorts of elaborate machines and contraptions. One of his mechanical devices trapped his wife Aphrodite—the goddess of Love herself—in the arms of her lover Ares, exposing her to the ridicule and judgment of the other Olympians. Technics is thus an integral part of the equation, alternately fusing and refusing the compulsion to make two from one, and vice versa.

Thus, despite all the radical fractures and mutations of the following two and a half millennia, this fundamental definition of love—a "meeting and melting into one another"—will remain essentially and remarkably intact. While European consciousness itself undergoes all sorts of upheavals (from the subtle withdrawals into the self at the beginning of the Christian era to the more obvious "modern discontinuities," such as Copernicus, Descartes, Darwin, and Freud), this basic portrait of love remains the West's Dorian Gray, albeit more prominently displayed than the painting in Wilde's story.

For instance, we can see strong echoes of Plato's atomic-fusional model in Georges Bataille's reading of eroticism in the middle of the twentieth century:

> Reproduction implies the existence of *discontinuous* beings.
>
> Beings which reproduce themselves are distinct from one another, and those reproduced are likewise distinct from each other, just as they are distinct from their parents. Each being is distinct from all others. . . . Between one being and another, there is a gulf, a discontinuity. It is a deep gulf, and I do not see how it can be done away with. None the less, we can experience its dizziness together. It can hypnotize us. This gulf is death in one sense, and death is vertiginous, death is hypnotizing.
>
> This gulf exists, for instance, between you, listening to me, and me, speaking to you. We are attempting to communicate, but no communication can abolish our fundamental difference. If you die, it is not my death. (1986, 12–13)

Indeed, it may even be argued that Lacan's notion of lack is but a complex supplement to *The Symposium*, particularly the often overlooked section dealing with gendered or sexuated desire.[6]

Moreover, as Aristophanes warns in the epigraph to this section (and in direct defiance of the etymology of the *indivisible* individual),[7] there is always the chance of being split *further still.* What we take to be a whole human being (i.e., one's *self*) is, mathematically and mythologically speaking, actually only a half. (Witness the phrase we use to introduce our partners: "She [or he] is my other half.") If we can assimilate this extremely long example of division, then perhaps we can be sliced again. In this case we would be only a quarter of the original One, and yet—once again—we could feasibly consider this quarter to be a functioning entity, albeit an entity which lacks.

The crucial aspect of this splitting effect is not so much the early echoes of a Deleuzian schizophrenic subject—one of his "dividuals"[8]—but the way in which such a mode intersects with the compulsion to count and record: "and that we shall be like tallies." Hence, when we speak of love, we are not discussing merely the primordial encounter between humans, which is only later coded through culture, but rather the *inherent instrumentality* which accompanies the emergence of the being that loves (or seeks love). Hephaestus's instruments have a healing power, but to fully fuse back to our prior androgynous state would be to revert to a prehuman, prehistorical condition. We can tally only those things which stand up to be counted. And before the emergence of people proper, the scarred and damaged Titans could not stand still long enough to be part of such a democratic process. They were too busy throwing thunderbolts and rolling along in blissful ignorance, à la Rousseau's Original Man.

Thus, for every Plato, who signifies the atomic model of love, there are a thousand potential Platos: one for every subsequent split in the subject, which, judging by the fable itself, will only recompose itself into a unit with yet more erotic valences. (If Aristophanes' basso-relievo figure is split again, he or she will necessarily have seven "true" partners to reestablish the unity of the One.) Such is the influence of the fable in the West that the alienation of the dating market, within the swarm of society itself, is a search for this original unity. The error, however, is the assumption that Zeus struck only once before our time and not repeatedly, as Aristophanes warned. Perhaps we have angered Zeus again since Plato's dinner party (an extremely likely situation, given our hubristic history). Thus, we fool ourselves in thinking that we are engaged in a search for the True Missing Piece, when we should be looking for several pieces. Or even the entire jigsaw puzzle.

Love in the postromantic world is thus a matter of exponentially proliferating valences and their control.

Libidinal Economies and Amorous Structures

> I couldn't have done it without you . . . or at least people very similar to you.
>
> —GEORGE SEGAL AS JACK GALLO, *Just Shoot Me!*

So much for the traditional genealogy. I now wish to focus on more recent readings of the lover's discourse, specifically those which begin to question the atomic-fusional model and the romantic-fatalistic recoding of Plato (i.e., the chivalric search for the destined lover). This entails paying attention to the *structure* of the love scene—in other words, the way in which the love narrative places us, interpellates us, and projects us into the libidinal economy.

From the moment we associate love with "having and holding" (as the marriage vow makes explicit), love is exposed as deeply imbricated in not only the wider symbolic economy but also the restricted exchange economy. In a chicken-and-egg dynamic, libidinal economies lead to, and emerge from, desiring circuits—that is, the way something called "love" circulates in our thoughts, our texts, and in society itself.

Let's think of some abstract transactions which fall within (an admittedly ahistorical) libidinal economy:

A man gives his neighboring village a cow, so they give him a woman.
A man gives a woman a compliment, so she gives him a kiss.
A man feels guilty about an affair, so he gives a woman some flowers.
A woman is angry at a man and decides to "freeze her assets" for the night.
A man and woman get married in order to—paraphrasing that hopeless romantic Immanuel Kant—have legal access to each other's genitals.
A man gets completely drunk, wondering why he made such a risky emotional *investment* in a woman who subsequently ran off with his brother.

While reading through these examples, it is quite likely that the reader has the same man and woman in mind. One scenario logically leads to the other, following the trajectory of a love *story*. Certainly, one situation seems to flow quite smoothly into the other, and this itself bears witness to the power of narrative in the lover's discourse—or more accurately, to the willful habit to organize events according to a specific narrative logic. To fall in love means to become hypersensitive to the signs being emitted by the beloved. Does she really love me? Is he sincere when he says that? The lover locates him- or herself in the structure, acting according to the code, assessing his or her personal *value* for not only the Other but for the libidinal economy itself.

The etymology of economy is in fact "keeping one's house in order." So, to speak of a libidinal economy is something of a tautology, since the notion of regulating the domestic sphere is at the heart of all economy. Indeed, it is particular to our notion of love that it both perpetuates capitalism (through marriage, conjugal property transmission, the profits derived from the desiring media machine, and so on) and simultaneously operates as a challenge to it. As Niklas Luhmann says, "It is impossible in love to calculate the costs or weigh up the accounts, because both one's profits and one's losses are enjoyed; indeed, they serve to make one aware of love and to keep it alive" (1998, 67). We need only think of a situation where a woman leaves her wealthy partner for the romance of shared Bohemian poverty, in order to see the slippages between a libidinal economy and a capitalist one.[9]

And yet the logic of exchange, accumulation, and profit, whether financial or "spiritual," infuses both. In other words, we should be careful not to confuse the market economy with the libidinal economy. Only a vulgar neo-Marxist could believe that love is nothing more than a capitalist conspiracy to keep the mode of production hidden from view, if only because love's signifying system has emerged in such distinctive cultural, political, historical, and economic conditions. We are not *affected* by the economy; we are *infected* with it. Its logic extends into different, though interrelated, spheres. So, while Hollywood and other discursive factories present love as something outside of the inhuman logic of capitalism, it can in fact be viewed from the perspective of emotional economies: of profit, reward, and loss. Indeed, perhaps what Barthes calls "the signifying economy of the amorous subject" (1990, 6) is the only comprehensive way to view love.

All economies generate their own structures, and the libidinal economy is no exception. As the epigraph to this section admits, we are all eminently and ultimately replaceable. In contrast to the touchingly humanistic American immigration law in which a foreigner can be employed only if bringing a skill which no American possesses, social structures care little for fetishizing the idiosyncratic talents of the individual. When the boss thanks her staff for working particularly hard to meet a deadline and claims that she "couldn't have done it without you," she is really attempting to deny the unspoken understanding that "you" could refer as much to an ambitious temp worker as to the allegedly unique being that happens to respond to your name.

The same can be said for the libidinal economy, a point not lost on Roland Barthes, even during his most sentimental moments: "The structure has nothing to do with persons; hence (like a bureaucracy) it is terrible. It cannot be implored—I cannot say to it: 'Look how much better I am than H.' Inexorable, the structure replies: 'You are in the same place; hence you are H.' No one can plead against this structure" (1990, 130).

It is this *inherent interchangeability* which lies at the brutal heart of the lover's discourse. The fact that almost every text produced in its name insists otherwise only serves to highlight the power of denial needed to keep such knowledge at bay. Love songs, poems, and confessions doth protest too much when they insist on the unique character and qualities of the paramour. Consider, for example, the romantic comedy staple whereby a lover accidentally calls out the wrong name of his or her partner in the moment of passion. This situation is not so much embarrassing in the sense of an ethical slipup but rather a taboo acknowledgment of the very interchangeability of the love object.

Think also of how we can each adopt the Shakespearean sonnets or Beatles ballads to our own situations and loved ones, and subsequently to different people at different stages on life's way. Barthes: "Though each love is experienced as unique and though the subject rejects the notion of repeating it elsewhere later on, he sometimes discovers in himself a kind of diffusion of amorous desire; he then realizes he is doomed to wander until he dies, from love to love" (1990, 101). This is Eros's dirty little secret, and we must unpack its implications if we are to understand the willful blind spots of the discourse which frames it. The critical question then becomes: How do we love while *simultaneously acknowledging* the intrinsic interchangeability of ourselves—and of the Other? Why does it feel so special and unique, when it is the oldest script in the business?[10]

Nevertheless, and despite appearances to the contrary, our purpose is not to bleed love of its romantic magic—to knock down the fading sets and reveal the illusion—but rather to expose the paradox at its center. Why? To reconfigure intimacy according to rhizomatic principles so that *there is no center*. For as soon as something acts as an essential organizing principle, we have a systemic weakness or vulnerability. And this in itself leads to the false consciousness and hypocrisy of the contemporary love story.

This is because as soon as we feel the stirrings of love, we submit to the amorous structure. When reading a novel or watching a movie, we subconsciously recognize the thoroughly public circuits channeling the narrative, a phenomenon once known as Bovaryism. "The subject painfully identifies himself with some person (or character) who occupies the same position as himself in the amorous structure. . . . Identification is not a psychological process; it is a pure structural operation: I am the one who has the same place I have" (Barthes 1990, 129). This becomes particularly clear in the case of jealousy: "The person with whom I can in fact talk about the loved being is the person who loves that being as much as I do, the way I do," that person being "my symmetric partner, my rival, my competitor." Above all, "rivalry is a question of place" (65).

Hence the trauma of the second love.

It is a commonplace observation that we romanticize our first "true" love according to the ancient principles handed down by Plato and repeatedly regurgitated by the romantic inheritors of the image repertoire. Witness the common conviction that fate is playing a major part in the love life of each of us, and the ease with which we shelve our secular, rationalist principles at the point of accounting for the first intimate relationship (outside the family fold). Something resembling *destiny* is sifted from the banal details of daily life, so that the encounter and subsequent affair become the stuff of personal legend. "You are my other half," we whisper to our first love. "This was meant to be."

The trauma comes not so much with the first broken heart, which is indeed a shock to the system, but rather with the next time we whisper these words . . . *to a different person.*[11] How can the discourse itself assimilate this flagrant contradiction? What if many women have the same shoe size as Cinderella? How to explain this security breach in the code—made explicit by the tattoo bearing a person's name, which far outlast the passion which inspired it? Only when we whisper these questions to ourselves do we realize the radical interchangeability on the level of structure, although it is rarely acknowledged explicitly. (To do so would leave oneself open to accusations of bitterness and cynicism.) The formalistic principles of the lover's discourse reveal themselves as ironclad, for while we may be able to seduce someone sexually by saying, "You could be anybody, and so could I," we would be spectacularly unsuccessful on the level of love. (Recall Lola's nocturnal question to her boyfriend in the introduction: "Why me and not one of those other girls?")

One of the ironies of love is that we experience our relationship to be deeply personal, for indeed, what could be more personal than love? Yet, undoubtedly, love stems from impersonal biological drives and the code of language itself, a vast impersonal structure which precedes us. How can this *really* be personal? We are assigned a role by our culture, and we obediently play along with it. Slavoj Zizek goes so far as to label this obedience "the automatism of love," and according to his less-than-romantic description, falling in love is "set in motion when some contingent, ultimately indifferent, (libidinal) object finds itself occupying a pre-given fantasy-place" (1997, 39). Following the totalizing Lacanian system, love is thus nothing more than the product of a given set of conditions, with little consequence given to the trivial differences between the equally trivial biographical details of those who participate in its "mysteries." Modes of desire are simply byproducts of a socio-semiotic machine which is effectively the aggregate of those trapped within a certain libidinal and linguistic economy. As Colette Solers puts it, emphasizing *different* versions of the code, rather than its backwards compatability: "Neither the Greek philia, nor the courtly model, nor divine mystical love, nor classical passion can captivate our present-day modes of *jouissance*, and we are left with love without any model, constructed as a symptom and attended only by the unpredictable conjunction of the contingencies of encounters and the automatons of the unconscious" (47).

The point can be clarified further still if we return to Barthes. Recalling a painful memory, the amorous Frenchman describes the moment when he is stood up by a lover: "I take a role: I am *the one who is going to cry;* and I play this role for myself, and *it makes me cry*" (1990, 161). He goes on to say, "I make myself cry, in order to prove to myself that my grief is not an illusion: tears are signs, not expressions" (182). After the automatic articulation ("I love you"), the lover becomes caught up in the tailwind of this decision and inserts him- or herself within the image repertoire. Henceforth, the lover is at the mercy of the code's signifying system, and all subsequent emotional welfare is conducted under the metasign of mediation itself. Something considered "internal"—a private life—is exposed, externalized. This movement begins elsewhere, though. The movement is centripetal—not centrifugal, as the poets would have us believe.

We begin this book, therefore, with a de-subjectivized, de-psychologized concept of love. This is in order to look at love, as it were, from the outside. *Love Is E(x)ternal.* That is our motto.

At least for now.

Interpersonal Interpenetration

> Scientific discourse has engendered all sorts of instruments that we must, from our vantage point here, qualify as gadgets. You are now, infinitely more than you think, subjects of instruments that, from the microscope right down to the radiotelevision, are becoming the elements of your existence. You cannot currently even gauge the import of this, but it is nonetheless part of what I am calling scientific discourse, insofar as a discourse is what determines a form of social link.
>
> —JACQUES LACAN, *On Feminine Sexuality*

> Marriages are made in heaven and fall apart in the automobile.
>
> —NIKLAS LUHMANN, *Love as Passion*

German theorist Niklas Luhmann treats love as a specific mode or vector of a wider cybernetic communications system. His book *Love as Passion* begins with the premise that "it is possible . . . to treat society as a social system that consists solely of communications and therefore as a system that can only reproduce communications by means of communications" (1998, 4). According to such a perspective, love is not so much an abstract emotional field, but constitutes "a symbolic code which shows how to communicate effectively in situations where this would otherwise appear improbable. The code thus encourages one to have the appropriate feelings" (8–9).

As a celebrated proponent of systems theory, Luhmann takes an exceedingly, and revealingly, abstract view of love as a discourse. Citing novels as the most significant of code-bearing technologies regarding the dissemination of love, Luhmann argues that our amorous vocabulary evolved in its current form during the eighteenth century. Even the same words began to function in radically different ways and in unprecedented contexts. This innovation was due to the fact that *amour passion* was the only model conceptually flexible enough to mask the contradictions emerging in the breakdown from the essentially stratified system of the late Middle Ages and Renaissance to the primarily functional one of early modernity.

It may take us by surprise, then, when we read his claim that "it is most assuredly incorrect to characterize modern society as an impersonal mass society and leave it at that" (12). Indeed, he goes on to claim that "modern society . . . affords more opportunities both for impersonal and for more intensive personal relationships." Moreover:

> It is possible to speak in terms of an enhanced capacity for impersonal relationships, in that one can communicate in numerous areas with no risk of misunderstanding, even if one has no personal knowledge whatsoever of the people with whom one is talking, and can only 'size them up' by means of a few hurriedly noted role characteristics (policeman, salesman, switchboard operator). This is the case, moreover, because every individual action depends on innumerable others, the functions of which are not guaranteed by certain personality characteristics that can be known to the person who has to rely on them. Never before has a society exhibited such improbable, contingent dependencies, which can neither be held to be natural, nor interpreted solely on the basis of one's knowledge of other people. (13)

This relatively recent social development has enormous implications for the emergence of whateverbeing, but we'll return to these later. For now, we are interested in Luhmann's notion of "interpersonal interpenetration," in which love becomes a form of multipurpose "affect-management." In this age of hyper-machinic bureaucracy—in which personal freedom boils down to the freedom to be highly regulated by the state and other institutions—love is a code which allows us some valuable, but ultimately illusory, breathing space. To quote Luhmann: "The individual person needs the *difference* between a close world and a distant, impersonal one," since the psychic labor of individuation occurs in this very space opened up during the rational-enlightenment age.[12]

Several significant points should be noted as soon as we consider love as a "generalized symbolic medium of communication" (18). The first is that this code functions "according to the rules of which one can express, form and simulate feelings, deny them, impute them to others, and be prepared to face up to all the consequences which enacting such a communication may bring with it" (20). Another point is the way in which the semantic matrix enables "love to be learned, tokens of it to be interpreted and small signs of it to convey deep feelings; and it is the code which allows difference to be experienced and makes unrequited love equally exalting" (ibid.)—a theme tackled in the 1940s by Denis de Rougement.

This leads to a paradoxical position for the beloved:

> If the other person lays claim to possessing a world-constitutive individuality, then one has always already been allocated a place in that world and therefore is *ineluctably* faced with *the alternative of either affirming or rejecting the other's egocentric projection of the world.* This complementary role as the affirmer of a particular world is forced upon one, although at the same time that world projection is supposed to be unique, idiosyncratic and therefore cannot serve as a basis for consensus.[13] (21)

Luhmann depicts a process which exposes an asymmetrical fracture between the protagonists of the love scene. Indeed, this fracture between lover and loved is perhaps even more constitutive and originary than Lacan's claim that there is no sexual relation, since Lacan is speaking about the negotiation of difference from the position of sexuation. According to Lacan, a male-sexuated subject has no relation with the female-sexuated subject because of the tyrannical structure of the phallocentric symbolic. Luhmann, in contrast, presents an even more fundamental fissure, one at the level of ontology, between the alter (the beloved) and the ego (the lover).[14]

Again, surprisingly, such a gulf is not crossed by "better communication" or more legible content but through the code itself. In a McLuhanesque move—which also harks back to Socrates' belief that love is primarily an intermediary[15]—the medium metamorphoses into the message:

> Love is able to enhance communication by largely doing without any communication. It makes use primarily of indirect communication, relies on anticipation and on having already understood. And love can thus be damaged by explicit communication, by discreet questions and answers, because such openness would indicate that something had not been understood as a matter of course. (25)

Hence the prevalence of "eye language" between lovers and long conversations about absolutely nothing.

"Being in love," according to Luhmann, "means to internalize another person's subjectively systematized view of the world" (25–26). Various obstacles and blockages also serve to enhance love rather than to thwart it—a romantic motif at least as old as Tristan and Isolde, although given a modern twist alongside developing self-reflexivity. Love is thereby figured under the sign of the messianic-utopic, and like Agamben's coming community, it is essentially "something to come," a pure potentiality whose seductive power and rhetorical purpose would be diminished through the usual clumsy attempts at actualization.

Love (i.e., the knowledge and language of love) was transmitted at an exponentially increasing rate via novels after the invention of the printing press. "Already by the seventeenth century," writes Luhmann, "it was common knowledge that the lady had read novels and therefore knew the code" (31). The lover's discourse can thus be figured as an early technological virus, infecting each generation with a more sophisticated version of itself. Alternatively, it could be viewed as an early example of open source coding, accessible to the epochal tinkerings of programmers who wish to adapt its functionality toward new dispensations of marriage and market.[16] (For Kittler also, people can be conceived of as "relay stations" for the transmission of the affective code, since in order to be, "love needs intercessors, or mouth-pieces, or translators" [1990, 73].)

People, accordingly, must become as legible as these novels for love to flourish: "In order for intimate communication to take place, the persons involved must be individualized to such an extent that their behavior can be 'read' in a specific way: on the basis of a difference. And this difference is that of their own immediate interests and/or habits and what is done out of consideration for the other person or the relationship to that person" (Luhmann 1998, 34). Suddenly, "identity" becomes an issue, as the private sphere begins to inflate itself within the public like a balloon within a balloon. A whole index of compatibility is inscribed between considerations of passion versus friendship, wife versus mistress, or guardian versus gallant knight.

Luhmann reminds us that medieval love poems were based on social stratification rather than individuality and that personal characteristics were almost entirely absent from the novel up until the late seventeenth century. One hundred years later, however, the modern image repertoire begins to cut some more familiar grooves in the libidinal landscape:

> As long as love was thought of as an ideal, *a knowledge of the object's characteristics* was essential. In the field of paradoxical codification, love justified itself by means of imagination. Once the autonomy of intimate relations had finally been established and raised to the level of reflexion, it was possible to justify love simply by the inexplicable *fact that one loved.* (44)

Love proceeds by way of a "spiral of enhancement."

For Luhmann, "words divide more clearly than do bodies, turning difference into information" (71). According to the principles of information theory, love is not immune to the laws of entropy. "By laying claim to time, love destroys itself," he writes. Since "the code's switch from a natural to an imaginary basis exposes the love to temporal corrosion, an erosion that is actually faster than the natural decline of beauty would be. Subjectification and temporalization go hand in hand" (73).

If we view love as a cybernetic code, producing new forms of community, we can have a better understanding of the implicit injunction to remain constantly in love. To not be in love, or not desire to be in love, is to reject the need for validation of self-portrayal. Paradoxically enough, to reject love's interpersonal potentiality—through the nunnery, the monastery, or the bunker mentality of the post-passion couple—is to reject society itself, since society is stitched together by the semantic matrix of this master code.[17] This is why *rejecting* the lover's discourse is *part of* the lover's discourse, and even utterances to the contrary only emphasize the *obligation to be continually in love.*

Where things get increasingly sticky—and where the code threatens to crash—is in the conflict between a communal, agapic love and an interpersonal, erotic one. More specifically, the question centers on how an individual, with an essence, assimilates the psychic violence of the trauma of the second love,[18] since this transference threatens to demolish the very notion of essence as a quality intrinsic to the individual. The code is inconsistent when it demands monogamy and fidelity, "for the beauty of the beloved justifies inconstancy, because it is also to be found in others. Indeed," Luhmann insists, "the hallmarks of the code, and above all the imperative of excessiveness, and beauty's function as stimulation now all go against the grain of this claim for exclusiveness" (1998, 214).

Before the advent of interpersonal interpenetration, love could not cope with the notion of "losing oneself in the other" without forcing a traumatic multiplication of the self. Now the notion of private internal worlds allows us to have a phalanx of desires and requirements which can rarely be served by one person alone. Marriage thus rewires itself into a social machine allowing the "combined maximum of order and freedom," providing a "haven of monotony," in its attempt to curb the inflow of libidinal information.

But against the sovereignty-based sanctimoniousness of Anglo-American psychoanalysis, interpersonal interpenetration precedes the individual who falls in love. "Interpenetration cannot be intended, conveyed, demanded, reached by pact or ended. Human interpenetration means precisely that the other person, conceived of as the horizon representing his own experiences and actions, enables the lover to lead his life as a self which could not be realized in the absence of love" (128). In other words, the discourse of love is functioning most effectively when we labor under the illusion that we are having our cake and eating it too, when we are spiritually connected to the entire human race *through* the beloved as a kind of holographic metonymy.

In Eric Rohmer's *Chloe in the Afternoon* (1972), the narrator-protagonist, Frederic, confesses:

> "That's why I love the city. People pass by and vanish. They don't seem to grow old. What in my eyes gives so much value to the streets of Paris is the constant yet fleeting presence of these women whom I'm almost certain to never see again. It's enough that they're there, conscious of their charm. Happy to test its effect on me, as I test mine on them. In a silent agreement without even the most subtle smile or look. I feel their attraction without giving in to it, and it doesn't estrange

> me from Helene. . . . On the contrary: These passing beauties are simply an extension of my wife's beauty. They enrich her beauty and share it in return. She ensures the world's beauty. And vice versa. When I hold Helene in my arms, I hold all women."

In his ever popular *New York Trilogy* (1992), Paul Auster makes the same audacious gambit:

> By belonging to Sophie, I began to feel as though I belonged to everyone else as well. My true place in the world, it turned out, was somewhere beyond myself, and if that place was inside me, it was also unlocatable. This was the tiny hole between self and not-self, and for the first time in my life I saw this nowhere as the exact centre of the world. (232)

Of course, such epiphanies are as fragile and time-sensitive as seemingly eternal and undeniable at the time of conception.

As early as 1653, Charles Vion d'Alibray published a book entitled *L'Amour divisé: Discours academique. Où il est prouvé qu'on peut aimer plusieurs personnes en mesme temps egalement et parfaitement (Love Divided: An Academic Discourse. Where It Is Proven That It Is Possible to Love Several People, Equally and Perfectly, at the Same Time)*. Were Oprah's book club to consider a contemporary version of this thesis, it would be met with a horrified curiosity, followed by popular censure. Exclusivity and constancy are still held in high regard, despite the actions of those who consider such values as an ideal. The task is not to read this behavior through a moralistic grid but to consider the ways in which love, as a technological movement toward community, accounts for the structural relevance of the third term.

The Three-Body Problem

> Love, like all divine powers, is not truly exalted except in a trinity.
>
> —GABRIELE D'ANNUNZIO IN PRAZ, *The Romantic Agony*

> Why can't we go on as three?
>
> —DAVID CROSBY, "Triad"

> Why should materialists, as they are called, be indignant about the fact that I situate—and why shouldn't I—God as the third party in this business of human

love? Even materialists sometimes know a bit about the *ménage à trois,* don't they?

—JACQUES LACAN, *On Feminine Sexuality*

To an extent, this study represents an enquiry into the metaphysics of modern love. Before we tackle the metaphysics of love, though, we must first come to grips with the *physics* of love. After all, both the language of love and the language of physics rely on an intense and sustained description and analysis of bodies—especially the relationship *between* bodies. One way out of the stagnant dynamic of Self vs. Other, or Lover vs. Loved, is by reference to a third term. Astronomy gave physics its own version of this conundrum, known as the three-body problem, in which the challenge is to describe and predict the behavior of a third mass in relation to two (much larger) masses. Newtonian mathematical physics could cope with the movement of two bodies but not three, and indeed an argument could be made that Western metaphysics experienced the same problem, more sublimated than solved by Hegel's dialectic. And yet, as we have already seen, the third term was recognized as far back as Aristophanes' speech at Plato's banquet: "The sexes were not two as they are now, but originally three in number; there was man, woman, and the union of the two." (The Christian doctrine of the Trinity, of course, shares such a structure.)

According to Michel Maffesoli, infinity itself begins with the third person. One is unified, contained, and perfectly isolated. Two is trapped within the gravitational pull of one. (Two is thus one plus one: a repetition or a reflection.) Three, however, is the beginning of that exponential velocity toward infinity, a raising to the next power. Three breaks through the glass ceiling of two, and from then on, there is no resistance from one number to the next. Three contains within it all positive numbers higher than itself (according to laws more metaphorical than mathematical). "With the figure '3,' society is born and therefore sociology" (1996, 104). This is due to the fact that a certain triangulation is required to counter the geometric and social stagnation of the couple, as with Lacan's interpretation of the Borromean knot: "No two rings of string are knotted to each other. . . . It's only thanks to the third that they hang together" (1999, 124). Note, for instance, how Crusoe is still considered an "outcast" despite the presence of Friday (before even the central question of race). It is only when a third person sets foot on an island that the islanders cease to qualify as simply outcasts but as a seminal society (before even the central question of gender).

It is precisely this third term—enigmatic, formalistic, essentially without content—which pulls at the bloated Newtonian bodies of the current lover's discourse. In contrast to the astronomical model (in which the third term is usually an asteroid), the third term in love has a greater force of attraction than the bodies themselves, or at least a far more significant sphere of influence. The third term is most certainly a satellite, but one which threatens to enter the atmosphere of the couple and shatter the Edenic tranquility of their own discourse: the waitress who smiles too much, the coworker who helps too much, the threat which can incarnate itself through a multitude of phantasms but which is never limited to only one.

The horror of the third term lies in the slippage it provides. Love is, of course, potentially universal. Yet, in its monogamous aspect, it rules out all others (a paradox unpacked by the writings of Kierkegaard). Once you triangulate the line of the couple, there is very little to limit its progress toward pure seriality. If the lover does not stop at one exclusive partner, then it is possible to be caught in pure metalepsis, emptying each encounter of all significance.[19] The intrinsic interchangeability of any social structure is exposed in all its identity-crushing indifference—an exposure which society itself cannot endure and must mask through the code itself, since the individual needs to feel that he or she has intrinsic value in order to function. So, we see the fragility of a situation in which the code must constantly deny that which it produces, that being the desire to love . . . that being . . . and, indeed, that one, too.

René Girard was one of the first critics to focus on the importance of the third term for the lover's discourse, stating in his book *Deceit, Desire and the Novel,* "In the birth of desire, the third person is always present" (1988, 21). While Freud had already noted that at least six people were phantasmatically "present" in any couple's bedroom (the couple plus their spectral parents), Girard emphasizes the *social* geometry of libidinal motivation. This shift, however, certainly has a psychological dimension, for it reveals the external origin of our desires. That is, "for lack of an auto-suggestion from within" (5), we attempt to seduce certain people not because we find them irresistible, but because they are potential trophies both suggested by, and wrestled from, someone else. Thus, for Girard, "The object changes with each adventure but the triangle remains" (2)—that is, the triangle created by the lover, the beloved, and what he calls "the mediator," the actual source of desire for the lover. "A *vaniteux* will desire any object so long as he is convinced that it is already desired by another person whom he admires" (7), and in the case of Don Quixote, "[t]he mediator is imaginary but not the mediation" (4). As a consequence, the affective movement toward the love object is in fact a subtle detour toward the mediator him- or herself:

> The triangle is no *Gestalt*. The real structures are intersubjective. They cannot be localized anywhere; the triangle has no reality whatever; it is a systematic metaphor, systematically pursued. Because changes in size and shape do not destroy the identity of this figure . . . the diversity as well as the unity of the work can be simultaneously illustrated. The purpose and limitations of this structural geometry may become clearer through a reference to "structural models." The triangle is a model of a sort, or rather a whole family of models. But these models are not "mechanical" like those of Claude Lévi-Strauss. They always allude to the mystery, transparent yet opaque, of human relations. (2)

Girard, however, is less forthcoming about the process *behind* this process; that is, he does not ask where this "initial" admiration comes from. Is this primary relationship with the mediator also mediated? And if so, what does this mean for that arrangement endearingly referred to by one of my Spanish students as a "lovely triangle"?

As Luhmann maintained above, the inherent presence of mediation means that there "is no question here of looking for the usual difference between copy and original for the very good reason that there is no original" (1998, 73). The evolution of the novel, according to Girard, is profoundly bound up with this understanding, foregrounded by Stendhal, Dostoevsky, and Proust. For these writers "jealousy and envy, like hatred, are scarcely more than traditional names given to internal mediation" (1988, 12), not least because individuals have relinquished their romantic sovereignty and become merely "those who operate the system" (3).

Simply put, "[n]ovelistic genius begins with the collapse of the 'autonomous' self" (38). Hence the contemporary ironic-Byronic situation whereby people explicitly follow the example of allegedly unique role models in order to feel unique:

> Subjectivisms and objectivisms, romanticisms and realisms, individualisms and scientisms, idealisms and positivisms appear to be in opposition but are secretly in agreement to conceal the presence of the mediator. All these dogmas are the

> aesthetic or philosophic translation of world views peculiar to internal mediation. They all depend directly or indirectly on the lie of spontaneous desire. They all defend the same illusion of autonomy to which modern man is passionately devoted. (16)

And it is this persistence of the "illusion of autonomy" that constitutes the conservative miracle of our time, barring us from thinking through the current impasse of personal and political community. Indeed, the fact that we are "passionately devoted" to this illusion should give us a clue where to look for the reasons behind this impasse: in the realm of the passions.

For Girard the attempted exodus from Romanticism was prompted by certain chronotopic developments (although he did not use Bakhtin's terminology). The geometry of desire moved from a Euclidean to an Einsteinian (novelistic) space. This corresponds to the psychosocial radiowaves of human action, moving from Adam Smith's invisible hand, through Freud's blind drives, to Althusser's coercive ideology. And so: "The distance between Don Quixote and the petty bourgeois victim of advertising is not so great as romanticism would have us believe" (31).

Yet, we are not completely satisfied with a purely structuralist approach to these phenomena. Alongside the conceptually infinite *interchangeability* of love, we must acknowledge the existence of a stubborn *irreplaceability:* a finite singularity of the Other (as those who have mourned a loved one only know too well). Herein lies the tension of contemporary intimacy, the tension between this radical interchangeability and the simultaneous awareness of our/their profound unsubstitutability on the level of personal narrative. It could be anyone, and yet it is irredeemably (th)us.

Things look very different from ground level, as they do from above, and any maps we decide to make during the course of this book have to be supplemented, and challenged, by the fractures and foxholes of the terrain itself. The *topos* of love constantly brings us back to the notion of essence, especially as the term is defined by Deleuze in *Proust and Signs.* When we isolate an individual from the background of the mass of humanity, are we isolating him or her under the sign of a unique essence? Is this "cookie-cutter" aspect of love immanent to the discourse itself? In other words, in professing our love, are we at the same time claiming a particular property or quality which belongs to the inamorata and to nobody else? Is love, therefore (and this is the key question), an *essentializing* phenomenon?

As we have seen, Agamben's notion of a coming community is based on an "inessential commonality." If so, then how can we reserve a place for love within Agamben's concept? Must we reconfigure the notion of a coming community to fit the rhythms of the human heart? Or is it a matter of retailoring love itself to fit the vision? Or does the role of technology in contemporary society alter the very foundation of this question?

In order to address these important issues, however, we must first spend a little more time on what we mean by both *essence* and *a coming community.*

TWO

The Storable Future and the Stored Past

> How does one create a memory for the human animal? How does one go about to impress anything on that partly dull, partly flighty human intelligence—that incarnation of forgetfulness—so as to make it stick? As we might well imagine, the means used in solving this age-old problem have been far from delicate: in fact, there is perhaps nothing more terrible in man's earliest history than his mnemotechnics.
>
> —FRIEDRICH NIETZSCHE, *The Genealogy of Morals*

> The idea of time plays such a magic part in the matter.
>
> —HUMBERT HUMBERT, IN NABOKOV, *Lolita*

In the mid-1990s a new technology debuted on our screens. Known colloquially as Bullet Time,[1] it employs a full 360-degree ring of cameras which simultaneously capture an object and moment from all angles. This effectively "freezes time," so that the spectator can virtually move around a person spilling a glass of orange juice or (in the case of one Nike advertisement) a bicycle falling and-yet not falling. It suddenly became possible to freeze the exact moment of a glass exploding, each fragment caught perfectly in the timeless moment of combustion. Or of a woman throwing a pillow at her lover in a passionate rage, so that the viewer can almost walk around the scene in a virtual 3D tableaux. It suddenly seemed possible that Nicholson Baker's concept of the Fermata was being rendered on-screen as an experiment before being applied to Real Life.

This special effect tended to prompt something approaching the technological sublime in the viewer, at least for the first few exposures. It seemed that we had been granted a glimpse of the secret inner workings of life, a revelation via the artificial arrestation of something we normally take for granted: the flow of time. Indeed, these images seemed to be examples of Benjamin's "chips of messianic time," to be read and decoded in their pregnant silence. Predictably, the modern media soon overexploited this new "technique" (for how are we to distinguish between a new technology and a merely new assemblage of existing technologies?), with the result that it became so common as to move swiftly into the realm of cliché, erasing any connection this new visual mode once had with a graceful phenomenological shock.

Biologists tell us that when adrenaline floods the nervous system, the perception of time actually slows down. Evolutionary speaking, this allows us to have more time in which to escape marauding mammoths. Similarly, the brain chemicals of a fly are processed so quickly that our rolled-up newspaper coming toward it is perceived in slow motion. (Unfortunately, evolution didn't factor in airplane travel—so, when we are strapped into our seats during turbulence, every nerve-shattering second is elongated, like satanic chewing-gum.) Any attempt to distinguish "time itself" from "the perception of time" would therefore have to take into account the Einstein-Bergson-Deleuze constellation, which itself points to a galaxy of accounts which extract the subject from time—or at least consider time as something far more complex than "homogenous, empty time," in which the subject is said to move like a commuter through a tunnel.[2]

The daguerreotype was the first technology to actually capture the contours of "frozen time," as opposed to the pictorial representation of painting; the uncanny effect of preserving an image of the ephemeral has been well documented by writers such as Roland Barthes and Susan Sontag. The melancholy structure of somehow "cheating" time, of capturing a token of immortality, of course only serves to highlight the mortality of that which is photographed. A photograph merely attests to something as past, a fleeting punctum which we can never truly grasp, but only mourn. Bullet Time allows us, however, to virtually enter the frame, bringing us another step closer to the unspoken goal of capturing time in a medium which allows us to rewind, fast-forward, and pause our own lives.

Take, for instance, the case of Humbert Humbert. The narrator of Nabokov's intricate tale not only describes his mother as "very photogenic" but claims to possess a photographic memory, the former perhaps being a happy result of the latter (or even, perhaps, encouraging such a faculty).[3] Although she is killed at the fringes of Humbert's memory—by lightning at a picnic when he is age three—her image is the first of several female ghosts to haunt this erudite fetishist with a penchant for nymphets and "the Proustian theme."

Indeed, Humbert is rocked to his foundations when he is first confronted with the image of Lolita. In this scene the noumenal Lolita coincides perfectly with not only the retinal image of Lolita lounging in the backyard but also Humbert's childhood memory of Annabel:

> It was the same child—the same frail, honey-hued shoulders. . . . I recognized the tiny dark-brown mole on her side.[4] . . . The twenty-five years I had lived since then, tapered to a palpitating point, and vanished. I find it most difficult to express with adequate force that flash, that shiver, that impact of passionate recognition. (Nabokov 1991, 39)

It is this "passionate recognition" which enables us to reread *Lolita* through the unconventional lens of science fiction, specifically through the trope of time travel. Humbert seems to fall into the space of revelation—itself a kind of "resonance machine"—as if two chronological moments telescope so perfectly that the past rises up within the present, as with Proust's madeleine.[5] The catalytic shock in this case is not taste but a coup d'oeil, igniting the "total recall" of his preadolescent love affair with Annabel.

It is the uncanny fusion of all these elements which make Lolita—both the book and the character—a consummate study in mnemotechnics, refusing a notion of desire outside the often violent conjunction between the "soft" technologies of language, law, memory, and phantasm and the "hard" technologies of writing, photography, and cinema.

Humbert in fact believes that there are two kinds of visual memory:

> one when you skillfully recreate an image in the laboratory of your mind, with your eyes open (and then I see Annabel in such general terms as: "honey-colored skin," "thin arms," "brown bobbed hair," "long lashes," "big bright mouth"); and the other when you instantly evoke, with shut eyes, on the dark innerside of your eyelids, the objective, absolute optical replica of a beloved face, a little ghost in natural colors (and this is how I see Lolita). (11)

These two kinds of visual memory echo the phenomenological writings of Henri Bergson, who wrote over a century ago:

> Whenever we are trying to recover a recollection, to call up some period of our history, we become conscious of an act *sui generis* by which we detach ourselves from the present in order to replace ourselves, first in the past in general, then in a certain region of the past—a work of adjustment, something like the focusing of a camera. But our recollection still remains virtual; we simply prepare ourselves to receive it by adopting the appropriate attitude. Little by little it comes into view like a condensing cloud; from the virtual state it passes into the actual; and as its outlines become more distinct and its surface takes on colour, it tends to imitate perception. But it remains attached to the past by its deepest roots, and if, when once realized, it did not retain something of its original virtuality, if, being a present state, it were not also something which stands out distinct from the present, we should never know it for a memory. (1988, 171)

The "virtual" in this text—written in 1896, during the birth of cinema (the Lumière brothers held their very first public screening the year before)—points to a certain technology which dwells within perception, so that mnemonic faculty itself is a form of technology ("a work of adjustment like the focusing of a camera"). Human memory accordingly mimics an artifact only in its infancy, a pure vector of recall as yet unsullied by sentiment and nostalgia.

Later in the story, after Humbert has wrenched Lolita from her oblivious suburban life and into his feverish road trip, he watches her playing tennis at school in Beardsley College:

> I could have filmed her! I would have had her now with me, before my eyes, in the projection room of my pain and despair! . . . That I could have had all her strokes, all her enchantments, immortalized in segments of celluloid, makes me moan to-day with frustration. They would have been so much more than the snapshots I burned! (Nabokov 1991, 231–32)

In contrast to Bullet Time or still photography, *movement* seems to be the key to capturing the moment. According to Humbert, a simple super-8 camera would be sufficient to immortalize Lolita's grace and preserve forever that combination of elements which had him drenched in "an almost painful convulsion of beauty assimilation"—a kind of pop-Kantian sublime processed through a quasi Freudian libidinal economy.

But is it really this simple? Is it merely a matter of having the right equipment at the right time to capture the moment?

In "The Seducer's Diary" by Kierkegaard, Johannes writes: "It would be of real interest to me if it were possible to reproduce very accurately the conversations I have with Cordelia. But I easily perceive that it is an impossibility, for even if I managed to recollect every single word exchanged between us, it nevertheless is out of the question to reproduce the element of contemporaneity, which actually is the nerve in conversation, the surprise in the outburst, the passionateness, which is the life principle in conversation" (1987, 399).

And, presumably, in a game of tennis.

Contemporaneity is something which no technology can fully capture, and it is that remainder (an accursed share, perhaps) which will always exceed the reproduction[6]—unless, that is, we manage to find a way to "actually," rather than "virtually," relive the moment. Such technology may bear little resemblance to a machine which reproduces the past, but perhaps be a technique which allows us to (re)experience the past as present—something like Nietzsche's eternal return. This is also felt in Chris Marker's *La Jetée* (1962), a film entirely composed of a succession of stills except for one moving shot, of a woman blinking, which functions as the exact opposite of the Bullet Time effect.

In Hirokazu Koreeda's 1998 film *After Life* (*Wandāfuru raifu*), the dead are first taken to an old school building in which they are asked to choose one favorite memory from their recently passed lives, which they can take with them to the afterlife. It is the task of the dedicated staff, who themselves are dead, to recreate this memory, using extremely low-budget props and techniques (part of the charm of the film). One old gentleman has trouble selecting a particular memory, not because his life was so eventful but quite the opposite: because it was the life of a typical salaryman with an arranged marriage. In order to prompt this gentleman's memory, his case manager orders all 70 tapes of his life (one tape for each year), and the old fellow spends several days sifting through the raw footage of his personal biography.

The tone of this austerely sentimental film issues from "magic realism" and therefore does not explain the logistics behind the existence of this operation, neither the institution itself nor the method used to film every second of every person's life and store it in a warehouse, waiting for someone to view the rushes. The most pressing question, however, is why this diligent branch of the postmortal bureaucracy *reenact* these people's memories, rather than just edit it out of the footage they clearly already have. It would do the writer and director a disservice to suggest that this is an error in the narrative logic of the film, since the inconsistency is so obvious. Rather, this process points to nostalgia at one remove, a touching belief in simulacra, on the condition that the subject has an input toward the production of the memory scene. The paradox is clear: In creating the copy, you more fully invoke the original. The "aura" of the event is transferred or recreated in the reenactment.[7]

One wonders whether Humbert Humbert, after dying of a coronary thrombosis (i.e., a broken heart) in his cell while awaiting trial, would have been satisfied with such a shabby facsimile of his cherished memories. It seems extremely unlikely that Koreeda's unnamed office could find an adequate starlet to play Lolita, especially when the "real" Lolita would presumably be gathering dust in their own archives. (Humbert makes it clear that we are reading his confessions only after the death of the other protagonists.)

Remarkably, Humbert—in the lost years after being abandoned by Lolita—composes and even publishes

> an essay on 'Mimir and Memory,' in which I suggested among other things that seemed original and important . . . a theory of perceptual time based on the circulation of the blood and conceptually depending (to fill up this nutshell) on the mind's being conscious not only of matter but also of its own self, thus creating a continuous spanning of two points (the storable future and the stored past). (Nabokov 1991, 260)

Such a "continual spanning of two points" brings us back into Bergsonian territory, specifically the Bergson reinterpreted by Deleuze, in which the element B (Lolita) appears as a *contraction* of A (Annabel).[8] Hence, the Annabel-Lolita hybrid is not simply a morphing of images in Humbert's mind but the chronotopia of virtuality itself becoming actual. Just as human perception experiences the past and the present simultaneously (both the "current" and the "previous" frames of a film), the materiality of presence embodies that which is (no longer):

> A succession of instants does not constitute time any more than it causes it to disappear; it indicates only its constantly aborted moment of birth. Time is constituted only in the originary synthesis which operates on the repetition of instants. This synthesis contracts the successive independent instants into one another, thereby constituting the lived, or living, present. It is in this present that

> time is deployed. To it belong both the past and the future: the past in so far as the preceding instants are retained in the contraction; the future because its expectation is anticipated in this same contraction. . . . The present does not have to go outside itself in order to pass from past to future. (Deleuze 1994, 70–71)

In other words, we dwell on the tip of the cone, "with the acoustics of time, domed time" (Nabakov 1991, 236).[9]

In this notion, then, we find ourselves anticipating those mysterious Proustian essences which provide a "superior viewpoint" by superimposing the past and the present—or more accurately, by contracting the past in the present, since we are always-already in the past, even in the so-called present. It is, however, Humbert's phrase "storable future and stored past" which resonates most with the topics at hand: the deployment of mnemonic technologies in contemporary culture, and the way such technologies can simultaneously enhance and erode the affectivity of their own "storage capacity."[10]

I am personally unaware whether Nabokov read much Bergson, but given his astonishing knowledge of nineteenth- and twentieth-century literature and ideas, it seems likely that some basic Bergsonian concepts captured his attention, especially when we consider his obsession with the "free world of timelessness" (1991, xxi). In his autobiography, *Speak Memory* (alluded to in the scholarly introduction to the annotated *Lolita*), Nabokov spatializes time, describing it as a spherical prison "without exits." Elsewhere in the same text, the author-narrator states, "I do not believe in time. I like to fold my magic carpet, after use, in such a way as to superimpose one part of the pattern upon another" (xxix), a literary variation of Leibniz's fold:

> for every dimension presupposes a medium within which it can act, and if, in the spiral unwinding of things, space warps into something akin to time, and time, in its turn, warps into something akin to thought, then, surely, another dimension follows—a special Space maybe, not the old one, we trust, unless spirals become vicious circles again. (xxxiii)

How to locate and define this special Space? Is it the space of literature? The space of play? Does it delineate duration, or the fold itself?

At this point, we have introduced two conceptions of time which seem to be at odds: time as pure movement (a future which is constantly pouring into the past through the present), and time as a rapid series of frozen moments (perhaps even jumping from one to the other, sometimes much slower than 24 frames per second). It is this first notion, however, which produces the possibility of the second, the artificial production of the (inherent) spatiality of time. The implication is that if we properly unlock the post-Einsteinian secrets of time—say, via Bullet Time—then we see how space itself is a product of time (and perhaps even allow us to construct a means of "traveling" "through" "time" which is less mediated and less melancholic than the effect of celluloid on brain cells via optic nerves, what Humbert refers to as "retrievable time" [261]).

Sometimes Humbert is caught in the static: "I would like to describe her face, her ways—and I cannot, because my own desire for her blinds me when she is near. . . . If I close my eyes I see but an immobilized fraction of her, a cinematographic still" (44). Sometimes he is at the mercy of the kinetic: "I prayed we would never get to that store, but we did" (51). His love for Lolita can become so overwhelming that it becomes the static of a certain "frequency," the kind of aural static that moves, as when he makes a request to the printer to repeat the word Lolita until it fills the page—the proper name of an improper obsession.

In contrast to Alfred Hitchcock, whose corpulent physique became the delight of vigilant cameo spotters, *film itself* spools around the words of the Humbert-Nabokov amalgam, daring the reader to recognize the significance of this nubile technology and its seductive relationship to the leering, overeducated advances of the written word.[11] As we have seen, Humbert laments the fact that he has no material record of his beloved and that his memory is merely photographic, not cinematic: "[P]ity no film had recorded the curious pattern, the monogrammic linkage of our simultaneous or overlapping moves" (58).

In matter of fact, Lolita *had* been captured on film, specifically the highly illegal blue movies directed by Clare Quilty at his mansion. But it is extremely doubtful whether Humbert, despite his depravities, would cherish this particular footage, since it would only serve to release the more poisonous memories of his loss, including his ultimate defeat at the hands of his rival. As with the Koreeda film *After Life*, Humbert's very effort of re-creating a memory (say of Lolita playing tennis) serves as the "saving" or "salvaged" simulacra of that which is "re-membered," the etymology being particularly relevant, when we consider how many times Humbert recalls his nymphets via their particular body parts: limbs, freckles, hair, and more.

Lolita herself is described as "a modern child, an avid reader of movie magazines, an expert in dream-slow close-ups." She is so much so that Humbert hopes that her early and extended exposure to the lover's discourse—via the media—will adequately prepare Lolita for his initial advance, and that she "might not think it too strange" should he try to kiss her, since she has seen its likeness on the silver screen (49). This phenomenon produces a neat twist, in that the annexed, sexless kingdom of modern childhood—dated by some with the publication of *Mother Goose's Melodies* in 1760—"loses ground" to the urgencies of desire, again thanks to the media, one of the (double) agents of the construction of innocence.[12]

Ultimately, it is Humbert's attempt to "fix once for all the perilous magic of nymphets" which leads to the book itself and the flirtation with cinema which lies at the heart of the narrative's logic. Being a man of letters, however, he eventually puts his faith in "aurochs and angels, the secret of durable pigments, prophetic sonnets, the refuge of art" (309), a form of immortality compromised by the medium in which it dwells.

This leads us to the function of Lolita herself, specifically the lo-fidelity of her behavior toward Humbert and to the passing of her precious nymphic years, defined as that "enchanted island of time" between the ages of 9 and 14. While sitting on a park bench surrounded by nymphets, Humbert pleads to nobody in particular, "Let them play around me forever. Never grow up" (21). He treasures the fleeting nature of nymphets, yet he wants to preserve them, like one of Nabokov's butterflies pinned to a velvet board in Lausanne's Cantonal Museum of Zoology.[13]

It is the fluid movement of the aging process which provokes in Humbert a desire to control this particular flux.[14] The "continuous spanning" of two points—whether we map these coordinates as Annabel in 1920s versus Lolita in the 1950s, or as too-young Lolita versus too-old Lolita—bracket the ungraspable. For Humbert the passage from nymphet to irrelevancy is too slow for the human eye to capture, yet other faculties and organs are only too sensitive to this process. The crucial question then emerges: Would Humbert fixate on the nymphet if there were no risk of her evaporation? Is it the imminent (and immanent) dissolution of her nymphancy which makes the *presence* of these creatures so unbearable, as if they were carved of smoke and impossible to hold at all? (Hence the moment of "redemption" in the novel, when Humbert confesses his enduring love for the pregnant and post-nymphic Lolita.)

It is this paradox which is missed as soon as the reader is tempted to accuse Humbert of pedophilia. Lolita is a cusp-creature, not simply a child. And it is this elongated metamorphosis which fascinates. Lolita's gangly legs straddle childhood and maturity, the past and the present, and it is the libidinal search for this hinge that constitutes Humbert's crime. Time dilates, time contracts. His is an ontological affliction, and perhaps no less grave for it.

Lolita—and more generally the nymphet (as we shall see later)—thus signifies time itself. Lolita *is* time. (Just as *Lolita* is chronography.) Humbert is incapable of fixing her perilous magic "once and for all," since such a technology (remembering that language itself is a technology) would have to resolve the paradox of freezing time *within* time, something Bullet Time hints at but does not achieve. If time can be conceived as anti-Euclidean spatialized movement, as Deleuze maintains, then it seems impossible to simply freeze time.

In Philip K. Dick's novel *The World Jones Made* (1993), the protagonist is born with one foot in the future and one foot in the past, so nothing comes as a surprise. (Hence the notable lack of crying when he is born, since he is prepared for it.) Similarly, the Tralfalmadorians in Kurt Vonnegut's *Slaughterhouse Five* can see in four dimensions, simultaneously inhabiting past, present and future: "They pitied Earthlings for being able to see only three. They had many wonderful things to teach Earthlings, especially about time" (1991, 26). Humbert himself is well aware of such temporal options and twice discusses the possibility that life constantly branches and forks, as certain theories of quantum physics maintain. "Time moves ahead of our fancies," he writes. Moreover, this knowledge leads him to wonder whether the girls that he phantasmatically molests are actually affected by his desires: "In this wrought-iron world of criss-cross cause and effect, could it be that the hidden throb I stole from them did not affect *their* future?" (Nabokov 1991, 21).

In a crucial scene, more explicitly tied to technology (and discussed in more detail below), Humbert realizes that Lolita remains completely oblivious to the carnal component of an orgasmic tussle on the couch: "The child knew nothing. I had done nothing to her. And nothing prevented me from repeating a performance that affected her as little as if she were a photographic image rippling upon a screen and I a humble hunchback abusing myself in the dark" (62). In trying to "possess" Lolita with neither her consent nor her knowledge, Humbert attempts to have his cake (Lolita) and eat it too (Lolita's image). Yet, he will end up hungry, due to the contradictory logic of mediated time.[15]

The term which links the vision of Jones, the Tralfalmadorians, and Humbert is the notion of "forever," a figure as familiar as it is abstract. *Forever* suggests a stasis, an immortality, traced by the eye reflecting the phantasmatic object: "The word 'forever' referred only to my own passion, to the eternal Lolita as reflected in my blood" (65). If the medium is time and the message is Lolita (or Humbert's passion for Lolita, which amounts to the same thing, diegetically speaking), then the medium is most certainly the message—a message Humbert seems to be sending himself, rather than the ladies and gentlemen of the jury, whom he so often addresses.

His confession stems from a string of separate "first impressions"—on the lawn, after summer camp, pregnant and married—all of which unfold in the "very narrow human interval between two tiger heartbeats" (111). (Consider Bataille's famous observation that eroticism moves in time as a tiger moves in space.) These moments, along with the attempt to represent them, force Humbert to confront "the stark lucidity of a future recollection (you know—trying to see things as you will remember having seen them)" (86). They are the punctuation marks, the punctual quilting points, of his encounter with essence. A temporal truth is thus glimpsed within the articulate white noise of his obsession: *His* Lolita, even an aging Lolita, unlocks the essence of love-within-duration. An enduring love. (And in this context, we are in a better position to understand the retrospective jealousy of Lolita's mother, Charlotte Haze, who "desired me resuscitate all my loves so that she might make me insult them, and trample upon them, and revoke them apostately and totally, thus destroying my past" [79].)

Perhaps we can justify being jealous of the past, when we actually—literally—live in the past.[16]

A Spectator Is Haunting Europe

> Memory is not so all-enwrapping, Dream sooner or later betrays itself. If an Actor or a painted Portrait may represent a Personage no longer alive, might there not be other Modalities of Appearance, as well?
>
> —CHARLES MASON, in Thomas Pynchon, *Mason & Dixon*

Every lover is the first love in terms of the simultaneous operations of the brain.
—JOHN BROCKMAN, *Afterwords*

Much is made of Humbert's European pedigree and Lolita's crass American charm. Whether or not we are tempted to make an allegorical reading of their fraught relationship, it seems appropriate to consider how Marx's specter has broken loose from the cellars and now inhabits even the "little ghost in natural colors" of Lolita and her celluloid siblings.[17]

Derrida, via Hamlet, has unpacked the implications of time's being out of joint. The uncanny persistence of the past, he insists, is distilled in the figure of the ghost, suggesting that the contemporary moment is haunted by its inability to assimilate that which came before. This has nothing to do with "reconciliation," or laying old ghosts to rest so that we may proceed with a clear conscience. Even less is it a case of "working through," as opposed to "acting out." It is, rather, a figure for the inherent fracture within certain human conceptions of time, as well as the symbolic economy we share with the dead, whether it be dead people, objects, or ideologies.

According to Humbert Humbert, one must be "a creature of infinite melancholy" in order to appreciate the ghostly symbolic power of the nymphet. "You have to be an artist and a madman . . . with a bubble of hot poison in your loins and a super-voluptuous flame permanently aglow in your subtle spine" (Nabokov 1991, 17).[18] Indeed, it is significant that Humbert's melancholy plagues him both before *and* after his encounter with Lolita and leads to several extended periods inside psychiatric institutions. Melancholy has traditionally been associated with Saturn, the god of time, the very same god who eats his own children. The erotic component of melancholy is thus a kind of preemptive and romantic loss, figured through both the spectral and the spectacle. The inaccessible or forbidden object becomes integrated in the subject via an introjection of the libido. The melancholic thus "keeps his or her own desire fixed on the inaccessible" (Agamben 1993b, 14).

Both Nabokov and Humbert play with the motifs of what Giorgio Agamben calls "this erotic constellation of melancholy." The latter's remarkable book *Stanzas* demonstrates how the love object is neither appropriated nor lost, but both possessed and lost at the same time, since "melancholy appears essentially as an erotic process engaged in an ambiguous commerce with phantasms" (1993b, 24).[19] The lessons of melancholy are thus that "only what is ungraspable can truly be grasped" (26) and that love itself is a form of "melancholic diligence" (27):

> No longer a phantasm and not yet a sign, the unreal object of melancholy introjection opens a space that is neither the hallucinated oneiric scene of the phantasms nor the indifferent world of natural objects. In this intermediate epiphanic place, located in the no-man's-land between narcissistic self-love and external object-choice, the creations of human culture will be situated one day, the interweaving of symbolic forms and textual practices through which man enters in contact with a world that is nearer to him than any other and from which depend, more directly than from physical nature, his unhappiness and his misfortune. (25)

The question then must be asked: Who possesses whom when Humbert and Lolita get in between the sheets of Nabokov's elliptical book? Loss haunts the former, no matter how many times he may have "had" her. Lolita is so much more—and *so much less*—than her body. On the one hand, she is the intangible and elusive spirit of the nymphet, representing an essence which Humbert cannot ever really grasp.[20] On the other, she is the trans-temporal phantasm of Humbert's desires, mocked by the "real" flesh-and-blood Lolita who chews gum, does soda burps, and gets bored. Humbert's Lolita never manages to overlap Lolita's Lolita.

Hence, "melancholia offers the paradox of an intention to mourn that precedes and anticipates the loss of the object" (Agamben 1993b, 20).[21] It is part of "the rigorously phantasmatic character of the amorous experience" (106) which leads to this "epiphany of the unattainable" (38). According to Agamben, then, the topos of melancholy traces a circle in which "the phantasm generates desire, desire is translated into words, and the word defines a space wherein the appropriation of what could otherwise not be appropriated or enjoyed is possible" (129). While Humbert holds the little ghost of Lolita in his arms, he is only too aware that the clock is ticking: "I knew I had fallen in love with Lolita forever; but I also knew she would not be forever Lolita. She would be thirteen on January 1" (Nabokov 1991, 65).

Humbert's frustration with the inability to fix Lolita as a moving image leads to the extended lament of spilled ink that makes up the narrative. And yet, as we have seen, it is this very frustration which produces the desire, according to the circular logic of melancholy, an intrinsic aspect of the lover's discourse. "The object of love is in fact a phantasm," writes Agamben, "but this phantasm is a 'spirit,' inserted, as such, in a pneumatic circle in which the limits separating internal and external, corporeal and incorporeal, desire and its object, are abolished" (1993b, 108). Thus, Humbert swings violently between the poles of the spectral and the material, at one moment even wishing to turn Lolita inside out "and apply voracious lips to her young matrix, her unknown heart, her nacreous liver, the sea-grapes of her lungs, her comely twin kidneys" (Nabokov 1991, 165)—the hidden signs of her ontological (in)accessibility.

As anyone who has played eye hockey with a fellow subway traveler knows, love is a disease of the eye (Agamben 1993b, 87). Rare is the Western tale of romance which begins with a perfume or song. Both Agamben and Nabokov attest to the significant and signifying role of optic interpenetration, the former noting that the whole cognitive process of Eros "is conceived as speculation in the strict sense, a reflection of phantasms from mirror to mirror. The eyes and the sense are both mirror and watcher that reflect the form of the object, but phantasy is also speculation, which 'imagines' the phantasms in the absence of the object" (81). (A dynamic circuit exploited as much by the adman as by the beloved.)

Through writing, Humbert seeks to turn the mirror of Lolita into a window, a window in which he can reexperience his past as present. "The basic drive in the human subject," states Silverman, "is the urge to see once more what has been seen before" (2000, 78). Despite the overwhelming impression of his rhetoric, Humbert's passion is not exclusive to Lolita but rather a vector affording access to essence. Through the time machine of confession (itself an interesting technology, reconfigured by Augustine, Rousseau, and Foucault), Lolita reanimates Annabel and, through Annabel, reanimates Humbert before the Fall—which, of course, is the fall into time. Thus, the specter haunting Europe turns out to be the spectacle itself, that being the scopophilic logic which traverses the sociopolitical history of the West. The aesthetic project of modernity, as embodied in the paradox of Humbert's unfocused intensity, marks the saturation point of this particular economy.

Despite the overlapping aporias, contradictions, and fissures in his account, Humbert's *fixation on fixation* points to an emerging awareness of Agamben's *whateverbeing*. The ontological slippages between Annabel, Lolita, Lolita's handmaidens, Lolita's mother, and of course Humbert himself allow both the narrator and the reader to resituate the unresolved tension between the radical interchangeability of the beloved and the profound irreplaceability of the same. Humbert writes, "And what is most singular is that she, *this* Lolita, *my* Lolita, has individualized the writer's ancient lust, so that above and over everything there is—Lolita" (Nabokov 1991, 45).[22]

We must, however, zoom in a little closer in order to answer the subsequent question: Who exactly is "*this* Lolita, *my* Lolita"?

That Complex Ghost

> It's a poor sort of memory that only works backwards.
>
> —THE WHITE QUEEN IN LEWIS CARROLL, *Through the Looking-Glass*

> Reckoned chronologically, this is correct. Thought historically, it does not hit upon the truth.
>
> —MARTIN HEIDEGGER, "The Question Concerning Technology"

"Did she have a precursor? She did, indeed she did."

So says Humbert Humbert. In fact, our narrator goes on to admit that "in a certain magic and fateful way Lolita began with Annabel" (Nabokov 1991, 9, 14). How do we read this claim in relation to whateverbeing, or the kind of *transdividualism* discussed thus far? How exactly does one person "begin" with an-other, outside the process of biological reproduction or allegorical Freudian blurring?

We have already mentioned the "passionate recognition" Humbert felt when he first saw Lolita. In this scene the phantasms which had haunted Humbert since his seaside romance with Annabel suddenly found a new object in which they could be transferred—hence Humbert's claim that he "broke her spell by incarnating her in another" (15). Such a reading, however, is too dependent on inadequate or distracting Freudian coordinates. Moreover, it becomes imperative to utilize certain concepts—such as narcissism, transference, desire, trauma—outside the strict Freudian constellation, accessing their pre- or post-Freudian resonance, specifically through ontology, phenomenology, and indeed Lacanian reconfigurations of the same.

We have already introduced the notion of the trauma of the second love. One method employed in order to avoid this trauma is simply to deny the ontological difference between two love objects, to fuse them into a hybrid. If Lolita is a miraculous incarnation of Annabel, then the trauma can be resolved, since there is no real transference, just an extension or expansion. While Humbert once loved Annabel, he now he loves Annabel *through* Lolita (and vice versa). But this answers nothing until we acknowledge how slippery such names function as signifiers of desire and designation.

We can see the stakes more clearly by returning to the quasi sex scene on the couch, in which Humbert wrestles with Lolita, up to and including the point of his own surreptitious orgasm. "Blessed be the Lord," remarks Humbert, "she had noticed nothing!" (61)

> I entered a plane of being where nothing mattered, save the infusion of joy brewed within my body. What had begun as a delicious distension of my innermost roots became a glowing tingle which now had reached that state of absolute security, confidence and reliance not found elsewhere in conscious life. With the deep hot sweetness thus established and well on its way to the ultimate convulsion, I felt I could slow down in order to prolong the glow. *Lolita had been safely solipsized.* (60, my emphasis)

On the one hand, we are presented with a lecherous middle-aged man, waxing and rationalizing the forbidden, yet ultimately banal, pleasures of domestic frottage. Yet, on the other, we have a remarkable concept, clouded in obscurity. "Lolita had been safely solipsized." From what? From Humbert's own disreputable desires? Beyond the moral taint of his own subjective *jouissance?* Or is the reference to Lolita-as-phantasm, wrapped in a bundle and incorporated into Humbert's fantasy world, in a process similar to, and perhaps preempting mourning, with all its cannibalistic and melancholic overtones?

In using *solipsize* as a transitive verb, Humbert both reduces Lolita to a doll-like figurine with which he can have furtive phantasmatic sex, and reduces *himself* to the kind of person who would find joy in such a process. When dealing with a nymphet, Humbert finds that he must throw out phantasmatic webs in which to catch his prey, and devour her—paradoxically—only after she "escapes" (at least until he transgresses the law completely). This is in direct contrast to his sexual encounters with Lolita's mother, in which he "possesses" Charlotte Haze physically yet refuses to devour her, or be devoured by her, even in the immediacy of the conjugal bedroom.

"I felt proud of myself," states Humbert, reflecting on his erotic tussle with Lolita:

> I had stolen the honey spasm without impairing the morals of a minor. Absolutely no harm done. The conjurer had poured milk, molasses, foaming champagne into a young lady's new white purse; and lo, the purse was intact.[23] Thus had I delicately constructed my ignoble, ardent, sinful dream; and still Lolita was safe—and I was safe. What I had madly possessed was not she, but my own creation, another, fanciful Lolita—perhaps more real than Lolita; overlapping, encasing her; floating between me and her, and having no will, no consciousness—indeed, no life of her own. (62)

This is a remarkable statement from Humbert, especially considering his stated investment in fixing the perilous magic of nymphets. Humbert's own phantasmatic artifact (this "fanciful" Lolita) floats between their two bodies, *perhaps more real than Lolita* herself. Here we are given a glimpse of the machinic nature of desire (or even love "itself"), in which a couple necessarily unfolds into an orgiastic multiplicity of partners and narcissistic reflections. We need only think of Freud's bedroom, crowded with the ghosts of parents, friends, and siblings; or of the plague of fantasies unleashed in Kubrick's *Eyes Wide Shut* (1999), leading to a literal orgy no less crowded than the Harfords' enormous Manhattan apartment.[24]

Humbert's project is thus encapsulated in the possessive pronoun, and he is continually at pains to distinguish *his* Lolita from other lolitas: "And what is most singular is that she, *this* Lolita, *my* Lolita, has individualized the writer's ancient lust, so that above and over everything there is—Lolita" (45). Again we see the trope of singularity versus multiplicity. This Lolita as opposed to what? Or whom? *That* Lolita? *Other* lolitas? All Lolita's "handmaidens," who, from Humbert's perspective, circle around his beloved like satellites around the sun?

At this point, then, we have identified at least three Lolitas: the imaginary (nymphet), the symbolic (schoolgirl), and the "real" (ungraspable, even by her). The narrative explicitly flags this polyontic quality when Humbert's chess partner, Gaston, asks, "*Et toutes vos fillettes, elle vont bien?*" Here Humbert realizes that Gaston "had multiplied my unique Lolita by the number of sartorial categories his downcast moody eye had glimpsed during a whole series of her appearances: blue jeans, a skirt, shorts, a quilted robe" (183). Thus, for Gaston there is a different Lolita for each of her outfits, just as for Humbert there is a different Lolita for each nervous heartbeat, a mental flip book at 24 frames per second, giving the illusion of simple continuity.[25]

Indeed, it is precisely this paradox between essentializing and fetishizing the singularity of the beloved, *when framed by the plurality of others*, that holds Humbert's (and indeed the lover's) discourse together. While he may dream of being shipwrecked on a desert island alone with Lolita, such a scenario attests only to the artificiality of such a situation, in which there is no third term (options, potentialities, threats) and which therefore cannot survive except in the circumscribed theater of fantasy.

But let us step outside literature and philosophy for a moment in order to find other examples that help underline this phenomenon. Consider a pop star and his or her role in the standard performative formation. Whether it be Madonna, Michael Jackson, or the latest star to hit the charts, the star dances in front, while the backup dancers shimmy behind and around him or her, not to be noticed in and of themselves but to delineate the contours of stardom itself. If there were not these anonymous supports behind the star, acting as the biomass from which the star is forged and distinguished, then the star would simply be extinguished for lack of oxygen. One cannot be singular without multiplicity. The backup dancer is like the character actor, a figure of whateverbeing, whom it would be absurd to fetishize or even notice according to the current economy of media identity cults. (At least since the Monkees, it is irrelevant whether this identity is "manufactured" or not, so long as it functions *as* identity.) Imagine Jennifer Lopez dancing on her own. It simply would not have the same impact without the "relief" of an ornamental humanity which provides the logic of distinction—what Heidegger calls "the ontological difference"—of a distinctive persona.

This same logic can be seen in Lolita's class list of names, a roll call which literally reads like a poem to the enamored Humbert. These thirty-nine students, listed with family name first, represent the sheer multiplicity of subjects as potential love objects, the miracle always being that we focus exclusively on one only if we are to satisfy the love code:[26]

> I am trying to analyze the spine-thrill of delight it gives me, this name among all those others. What is it that excites me almost to tears (hot, opalescent, thick tears that poets love to shed)? What is it? The tender anonymity of this name with its formal veil ("Dolores") and that abstract transposition of first name and surname, which is like a pair of new pale gloves or a mask? . . . Or is it because I can imagine so well the rest of the colorful classroom around my dolorous and

> hazy darling: Grace and her ripe pimples; Ginny and her lagging leg; Gordon, the haggard masturbator; Duncan, the foul-smelling clown; nail-biting Agnes; Viola, of the blackheads and the bouncing bust; pretty Rosaline; dark Mary Rose; adorable Stella, who has let strangers touch her; Ralph, who bullies and steals; Irving, for whom I am sorry. And there she is there, lost in the middle, gnawing a pencil, detested by teachers, all the boys' eyes on her hair and neck, *my* Lolita. (52–53)

The gulf separating a bureaucratic document and the living flesh of the narrator's desires is sustained by the increasing modernization of the lover's discourse. Anyone who has experienced an erotic shock at seeing his or her beloved's name in print, especially surrounded by other, far less significant names, understands the potency of this dynamic. (One unpublished short story I have read features a narrator who masturbates to his beloved's name in the phone book, a perfect illustration of this theme.) "Lolita"—real name Dolores Haze—is flanked by a Mary Rose Hamilton and a Rosaline Honeck and is thus "a fairy princess between her two maids of honour." (And much later Humbert will explicitly remind the reader "what importance I attached to having a bevy of page girls, consolation prize nymphets, around my Lolita" [190]).

Such logic, of course, is not exclusive to this century. We can in fact see the same impulse underpinning the following remarkable passage from Kierkegaard's "Seducer's Diary," in which the narrator, Johannes, implicitly considers the properly Scandinavian invention of the smorgasbord:

> What is glorious and divine about esthetics is that it is associated only with the beautiful; essentially it deals only with belles lettres and the fair sex. It can give me joy, it can joy my heart, to imagine the sun of womanhood sending out its rays in an infinite multiplicity, radiating into a confusion of languages, where each woman has a little share of the whole kingdom of womanhood, yet in such a way that the remainder found in her harmoniously forms around this point. In this sense, womanly beauty is infinitely divisible. But the specific share of beauty must be harmoniously controlled, for otherwise it has a disturbing effect, and one comes to think that nature intended something with this girl, but that nothing ever came of it. (1987, 428)

Johannes thus seems to acknowledge that each particular beauty emerges from the whole kingdom of womanhood; the aesthetic genetic pool, if you will. Accordingly, the singular girl which catches his eye becomes the only incarnation he can possibly imagine seducing (just as Cordelia becomes the object of his obsessive attentions):

> My eyes can never grow weary of quickly passing over this peripheral multiplicity, these radiating emanations of womanly beauty. Every particular point has its little share and yet is complete in itself, happy, joyous, beautiful. Each one has her own: the cheerful smile, the roguish glance, the yearning eye, the tilted head, the frolicsome disposition, the quiet sadness, the profound presentiment, the ominous depression, the earthly homesickness, the unshriven emotions, the beckoning brow, the questioning lips, the secretive forehead, the alluring curls, the concealing eyelashes, the heavenly pride, the earthly modesty, the angelic purity, the secret blush, the light step, the lovely buoyancy, the languorous posture, the longing dreaminess, the unaccountable sighing, the slender figure, the soft curves, the opulent bosom, the curving hips, the tiny feet, the elegant hands.
>
> Each one has her own, and the one does not have what the other has. When I have seen and seen again, observed and observed again, the multiplicity of this world, when I have smiled, sighed, flattered, threatened, desired, tempted, laughed, cried, hoped, feared, won, lost,—then I fold up the fan, then what is scattered gathers itself together into a unity, the parts into a whole. Then my soul rejoices, my heart pounds, passion is aroused. This one girl, the one and only in all the world, she must belong to me; she must be mine. (428–29)

It is worth dwelling on this passage, since Johannes has traced something crucial to the lover's discourse and, therefore, to the related constellation of community and ethics. "This peripheral multiplicity"—figured in the book by various cameo appearances by young maidens of the "fishmonger's daughter" variety—is a heterogeneous hypertext of qualities which splash across the entire gender, landing on some and not others.

"Each one has her own," suggests Johannes. And yet it is not merely a case of a unique essence, for he can "fold up the fan" and reterritorialize all these women under the universal signifier Woman, specifically in the body Cordelia, who becomes the representative of her sex. (Yet, if Johannes were to be pressed as to Cordelia's singular uniqueness, we may not get a straight answer.) In this metaphor of "folding the fan," then, Kierkegaard anticipates—and complicates—this point of Deleuze's: "It is not the subject that explains essence, rather it is essence that implicates, envelops, wraps itself up in the subject. Rather, in coiling round itself, it is essence that constitutes subjectivity. It is not the individuals who constitute the world, but the worlds enveloped, the essences that constitute the individuals. . . . Essence is not only individual, it *individualizes*" (2000, 43).

Returning to Lolita, it is clear that she also becomes individualized, if not completely solipsized, by essence. After marrying Lolita's mother, Humbert notes, "I kept telling myself, as I wielded my brand-new large-as-life wife, that biologically this was the nearest I could get to Lolita; that at Lolita's age, Lotte had been as desirable a schoolgirl as her daughter was, and as Lolita's daughter would be some day" (Nabokov 1991, 76).

Such a temporal notion concerns "the virtual"—not as in cyberspatial "virtual reality" but rather in the achronological unfolding of things through nonlinear time. It is in this sense of the virtual that the "nymphet" *always-already* qualifies as "woman." In simple terms, the nymphet *is* a virtual woman. What Humbert desires in Lolita is not exclusive to her and her alone but rather tied to a particular moment in the ontological chain of being. If the only thing Humbert loves in Lolita is her nymphancy, and her nymphancy is not essential to Lolita (since she will leave it behind in a few birthdays), then he does not love Lolita's essence. Through Humbert's eyes there is no essence to Lolita, to her individuality, since the ephemeral quality that attracts him can be passed on through the continuum of generations, as he notes while pondering Charlotte Haze:

> How different were her movements from those of my Lolita, when *she* used to visit me in her dear dirty blue jeans, smelling of orchards in nymphetland; awkward and fey, and dimly depraved, the lower buttons of her shirt unfastened. Let me tell you, however, something. Behind the brashness of little Haze, and the poise of big Haze, a trickle of shy life ran that tasted the same, that murmured the same. A great French doctor once told my father that in near relatives the faintest gastric gurgle has the same "voice." (92)

Another name for this "trickle of shy life" is, of course, essence.

After losing Lolita, Humbert is haunted by her in his dreams. However, she does not possess the kind of solid identity that we associate with our wakened state; rather, "she appeared there in strange and ludicrous disguises as Valeria or Charlotte, or a cross between them" (254). This "complex ghost" is a cipher or amalgam of his past wives and present obsessions, suggesting that the criteria we use to distinguish one lover from another—or indeed, simply one person from another—is suspended in the oneiric state. This familiar phenomenon points to the trickle of shy life which flows between all monads, and indeed slowly but patiently erodes the solid foundations of modern subjectivity.

If we leap out of the book, for the moment, into the "real" world, we also see how the character of Lolita is similarly an amalgam or synthesis. When asked about any research undertaken for his novel, Nabokov stated, "I travelled in school buses to listen to the talk of schoolgirls. I went to school on the pretext of placing our daughter. We have no daughter. For Lolita, I took one arm of a little girl who used to come to see Dmitri [his son], one kneecap of another," and thus a nymphet was born (xl).[27] In a clear case of the permeability between life and art, the whateverbeing of Lolita is lifted from the whateverbeing of several anonymous girls, who nevertheless inhabit the character, as essence, as much as the ghost of Annabel.[28]

As we have seen, Lolita shares the lease of her selfhood with several others, each of whom refracts still others in a prism effect. (Recall recent advertisements for "safe sex" which emphasize the interconnected nature of human collectivity, plainly stating that when you sleep with one person, you are simultaneously sleeping with the—perhaps viral—ghosts of his or her sexual biography.) Humbert is well aware of such fractal rhetoric and repeatedly addresses the various genealogies from which it emerges: "A breeze from wonderland had begun to affect my thoughts, and now they seemed couched in italics, as if the surface reflecting them were wrinkled by the phantasm of that breeze" (131). Indeed, in one particularly lucid moment, Humbert confesses that his lust for the "fiery phantasm" of nymphets may actually stem from the fact that there is "no possibility of attainment to spoil it by the awareness of an appended taboo" (264). For

> it may well be that the very attraction immaturity has for me lies not so much in the limpidity of pure young forbidden fairy child beauty as in the security of a situation where infinite perfections fill the gap between the little given and the great promised—the great rosegray never-to-be-had. (264)

This particular form of "security"—shattered the moment he wrenched Lolita from the phantasmatic or imaginary realm to the Real—refers directly to a kind of pure potentiality, one which became compromised as soon as it passed into the relatively degraded state of actuality (the legacy of Platonic idealism, perhaps, as much as the postromantic forces under discussion throughout this book). The twin engines of drive and desire compel Humbert to realize his fantasies in regard to Lolita, which catapults him into an ambivalent realm somehow "beyond happiness." To first be circumscribed by the law, and then actively prosecuted under its name, leaves few places to inhabit other than the potential. Thus, Humbert must swap the kinetic energy of the fugitive for the potential energy of the writer, at once a curse and a consolation.

Arachnography: Writing the Web

At one point in the novel, Humbert describes himself in the following terms:

> I am like one of those inflated pale spiders you see in old gardens. Sitting in the middle of a luminous web and giving little jerks to this or that strand. My web is spread all over the house as I listen from my chair where I sit like a wily wizard. Is Lo in her room? Gently I tug on the silk. She is not. (Nabokov 1991, 49)

Only a few pages later, he repeats the claim, referring to himself as Humbert the Wounded Spider (54). More than simply playing the frightening, hairy arachnid to Lolita's Miss Muffett, however, Humbert functions in the Deleuzian sense of a literary machine. In this case, Humbert's web is not the tool of his trade—à la Nabokov's butterfly net—but rather an extension of the organless body of the spider-machine itself: "The spider too sees nothing, perceives nothing, remembers nothing. . . . Without eyes, without nose, without mouth, she answers only to signs, the merest sign surging through her body and causing her to spring upon her prey" (182).

Humbert of course seems to do nothing *but* see, perceive, and remember. However, these operations are conducted in the mode of spider and not the traditionally human(ist) narrator. In this context perhaps it is not too violent or clumsy to transpose Deleuze's comments concerning Proust's narrator to Humbert himself—specifically, the claim that "the narrator has no organs or never has those he needs, those he wants" (2000, 181). This is due to the fact that "the web and the spider, the web and the body are *one and the same machine*" (182, my emphasis). Humbert thus extends his exoskeleton beyond the confines of his holding cell in order to catch the perilous magic of Lolita in the sticky threads of his prose.

Significantly, the Australian slang for a pedophile is *rock spider*, a name which acknowledges the arachnid's particular predator assemblage in relation to its prey. To put it in crude Deleuzian terms, a lion (in contrast) is a molar organism: It has *organs.* It assesses the world through a phenomenological matrix which prompts the human world to bestow it with a royal and dignified demeanor. Hers is an autonomous and organized mode of approach, ruled by the head and the heart. In contrast, the spider is a patient and ex-centric creature, relying on a strategy whereby it literally spins its own world from its own belly, making and unmaking its environment and thus itself. This is an altogether different "worldwide web" from the one we are used to, although we find a relevant resonance on the level of public concern over the internet as a "pedophile's paradise." The spider-as-organism may in fact be molar, but it is constantly becoming-something-else, through the various webs linking it with parasites, prey, and the various interlocking phyla of the ecosystem.

More specifically, the spider's web—Humbert's web—is an articulation of the technology of writing itself. It is a tangible, emergent property of the meaning-making machine, and its purpose is to read, capture, and extract. When Humbert *tells us* that he is testing his web for Lolita's exact location, he is both reading and writing signs.[29] Thus, to write *is* to read—and any attempt to separate these functions into different operations, albeit of the same process, is already working on an artificial, indeed impossible, conceptual division of labor.

The previous point can be made with recourse to the worldwide web itself, which, as all teachers now know, has become the world's largest, laziest, and least reliable library.[30] According to this now indispensable resource, two German scientists, Michael Stuke and Markus Koch, are literalizing the common metaphorical link between information technologies and spiderwebs. Working at the Max Planck Institute for Biophysical Chemistry in Göttingen, these researchers have managed to use silk from the black widow spider as the raw material for "nanowires"—ultrathin conductors which could "spark a revolution in miniature electronics" (Zandonella, 2001, 20). If their research fulfills its promise, fiber-optic cables could be a thing of the past, and our worldwide web may literally be connected by spider silk threads, cut to approximately one-twentieth of their normal diameter without any loss in strength or flexibility.

Harnessing "nature" for "culture" in this way (an instance of technology) only serves to blur the already tenuous and disingenuous distinction between the two realms. While Humbert is only a *metaphorical* spider according to all the topologies we are familiar with, he is engaged in an intense exercise of "becoming-spider" according to the fluxing categories proposed by Deleuze. Just as a workhorse is "closer" to an ox than a racehorse in Deleuze's system, Humbert is closer to a spider than another person who writes and lives according to a different ethical system. Ultimately, Humbert gets caught up in his own web, but this particular inevitability is less a result of some kind of poetic justice than a purely mechanical one. He lives *on* and *through* his web, so his fate unfolds according to the exigencies of this particular literary machine.

Nabokov states that the germ of his novel was provided by a news item concerning an ape held in captivity at the Paris Zoo. When encouraged to draw a picture, the ape drew the bars of his cage. In line with this genealogy, Humbert shares a simian heritage, as well. But the point is not to multiply and distribute Humbert along the Orwellian line of various animals—deciding whether he is now a spider and then an ape—but to emphasize the role of the "anthropological machine,"[31] of which "literature" comprises but one cog.

According to Zandonella, the scientists have also found that "spiders can learn from experience, with mature spiders building better webs than novices." Researchers from the Universities of Vienna and Melbourne—Astrid Heiling and Mariella Herberstein, respectively—have studied two species of orb web spider, *Argiope keyserlingi* and *Larinioides sclopetarius*, and concluded:

> In both species, practised builders made bottom-heavy webs compared with the novices' more symmetrical attempts, because it is quicker to run down from the hub, rather than up, to pounce on prey. 'The use of cumulative experience has not been shown [in spiders] before,' says Heiling. (2001, 20)

Such research seems to reflexively trace the Darwinian logic which lies dormant within traditional literary theory, especially when considered alongside Lacan's belief that spiderwebs allow us to "grasp the limits, impasses, and dead ends that show the real acceding to the symbolic" (1999, 93). As a consequence, the relationship between spider to web, author to text, should not be seen purely in terms of instrumentality or tools but rather a symbiotic form of bringing the world into being.[32]

During his famous lecture series on Hegel, Alexandre Kojève noted, "It is necessary to admit that after the end of History, men [will] construct their edifices and their works of art in the same way that the birds construct their nests and the spiders weave their webs" (Agamben 2004, 9). In his book *The Open: Man and Animal*, Agamben dwells on an interesting detail: "The spider knows nothing about the fly, nor can it measure its client as a tailor does before sewing his suit. And yet it determines the length of the stitches in its web according to the dimensions of the fly's body, and it adjusts the resistance of the threads in exact proportion to the force of impact of the fly's body in flight" (2004, 41–42). And thus, despite the fact that these two creatures are intimately connected by the choreographies of ecology, the "two perceptual worlds of the fly and the spider are *absolutely uncommunicating*" (42, my emphasis). It is at this point that the parallel with Humbert becomes clear, as witnessed in the scene where our narrator offers Lolita "a penny for your thoughts." Her response is to mutely stretch out her hand, in an attempt to simultaneously literalize—and therefore negate—the transaction. That is to say, the spiderweb, for Humbert as much as for Agamben, expresses the "paradoxical coincidence of . . . reciprocal blindness" (Agamben 2004, 42).

In his study *The Extended Organism*, biophysicist Scott Turner suggests that "animal structures like webs, nests, hives, burrows and mats are physiological extensions of their animal creators. They take energy and materials from the environment, including sunlight, water and oxygen, and funnel them to the organisms inside. This makes these structures as much a part of a living animal as more conventional organs such as livers, lungs, kidneys and hearts" (in Brown 2000, 30). From this perspective "the boundary between the living and non-living seems very arbitrary."[33]

In relation to this constellation—mapped between extended phenotypes, animal architecture, and malleable environments—the key question becomes: What is the difference between architecture and a stable exoskeleton? If a termite mound is "really an extension of the termites themselves, an integral and active part of their physiology" (31), then the skin or skeleton may be one of the most arbitrary limits ever enforced, based as it is simply on genetics, rather than a more holistic biophysical understanding.

In keeping with our discussion, Turner offers the example of the diving bell spider, which manages to live underwater due to the construction of an external "lung," basically a web that traps a bubble of air below the surface. Moreover, other species of spider have managed to acquire the ability to create dummy, or decoy, spiders inside their webs in order to fool any nearby predators. "Nature is full of energy ready to be tapped," states Turner, "and I see engines everywhere" (32). Learning how to adapt to the assemblage—or how the assemblage adapts *us* to its mobile structure—becomes inscribed in the expanded definition of literature which must surely accompany the recent developments in technology and media.

Humbert's arachnography can indeed be viewed as a form of "emergent literature"—specifically, symbolic activity with real-world effects. According to Wlad Godzich:

> The term "emergent" does not belong to the discourse of economics, but to that of evolutionary biology. Far from positioning a single, prescribed or even described path of evolution, it refers to what biologists call an emergence, namely the appearance of some functional features that are unforeseeable from the path of evolution of the organism concerned. In other words, the word "emergent" means exactly the opposite of the word emerging. (2000, 3)

Following on from this important distinction, Godzich states: "We need a theory of biological semiotics and a theory of creation semiotics if we are to build devices which construct their own semantic relations to the world. . . . In effect we must consider [literature] . . . as a creative intelligence that co-evolves sense receptors, computational coordinators and effectors needed for specific tasks" (2000, 6, 8)—a supplementary perspective on Deleuze's literary spider-machine.[34]

In contrast to this, the legacy of literary criticism has been either to focus on authorial expression, intent, and even "genius," or to play the glass bead game of semantic formalism within truth value systems. Like E. B. White's spider Charlotte, writing is compelled by the ethical or satirical—the author is "commenting" on society or even hoping to make an instrumental intervention in a certain debate or event. "Terrific," she writes in her web—the implicit rhetorical counterpoint to every editorial and the stock phrase of every literary critic. Literature is thus kept in check by those who wish to read it only in the service of a moral lesson or some other anthropocentric, demonstrative purpose. Accordingly, a story without redemption or revelation is considered no story at all, or at least an incomplete one.

It should not surprise us, then, that one of the most common readings of Nabokov's novel is a tale of ultimate redemption, since Humbert professes his love for Lolita even when she has long departed that "enchanted island of time" which previously represented the sole condition for such intense affect:[35]

> I looked and looked at her, and knew as clearly as I know I am to die, that I loved her more than anything I had ever seen or imagined on earth, or hoped for anywhere else. She was only the faint violet whiff and dead leaf echo of the nymphet I had rolled myself upon with such cries in the past . . . but thank God it was not that echo alone I worshiped. What I used to pamper among the tangled vines of my heart, *mon grand péché radieux*, had dwindled to its essence: sterile and selfish vice, all *that* I canceled and cursed. You may jeer at me, and threaten to clear the court, but until I am gagged and half-throttled, I will shout my poor truth. I insist the world know how much I loved my Lolita, *this* Lolita, pale and polluted, and big with another's child, but still gray-eyed, still sooty-lashed, still auburn and almond, still Carmencita, still mine. (Nabokov 1991, 277–78)

Thus, what for many critics is the moment of redemption—Humbert's love *in duration*, an *enduring* love—can alternatively be seen as the moment he succumbs to the overdetermined code of the lover's discourse. ("What I used to pamper . . . had dwindled to its essence.") Lolita no longer *is* an essence, but rather *has* an essence—one which has become identifiable, like a stone polished smooth in the clear river of time. This essence is a property of Lolita, rather than the other way around. Essentializing Lolita thus, individualizing her beyond all reasonable doubt, is the necessary moment for the novel to qualify as a love story ("the *only* convincing love story of the twentieth century," according to *Vanity Fair*). However, this means that Lolita herself (and now we *can* securely use such a term as "herself") is no longer a candidate for whateverbeing, at least through the eyes of Humbert, since her qualities are bound by her body and no longer overlap the phantasms of alterity.

Thus, we have seen how the "virtual" (defined as the potential achronological unfolding of things) circulates throughout the text in order to collect at this particularly dense point in the narrative. Just as intelligent spiderwebs are bottom-heavy in order to funnel the force of gravity toward the prey, Humbert's web draws the narrative toward the chronotopia of epiphany. Modern love thus flirts with the metaphysics of a coming community but is ultimately reterritorialized by the essentializing code. Humbert really does love *this* Lolita, and only this one. And so, there is nothing left to do other than die in isolation of a broken heart. (Or retire to a hotel in Montreaux and live off royalties.)

THREE

In the Artificial Gardens of Eden-Olympia

All this alienation . . . I could get easily used to it.

—JANE IN J. G. BALLARD, *Super-Cannes*

J. G. Ballard's business park for the technocratic elite, Eden-Olympia, has its own grinning Cheshire cat, a psychiatrist by the name of Dr. Wilder Penrose, whose "grimace of pleasure seemed to migrate around his face, colonizing new areas of amiability" (171). Penrose oversees the general well-being of the talented and highly paid workaholics who comprise the population of this "virtual city conjured into the pine-scented air like a *son-et-lumière* vision of a new Versailles" (8)—a "humane version of Corbusier's radiant city" (5). As such, Penrose takes it upon himself to engineer a suitable set of living arrangements for this highly concentrated form of postmodern community, whose citizens are "an order of computer-literate nuns, committed to the sanctity of the workstation and the pieties of the spreadsheet" (8), as well as willing participants in "a huge experiment in how to hothouse the future" (15).

The corporate wonderland of Eden-Olympia also has its own Alice, in this case a middle-aged Englishman by the name of Paul Sinclair. Paul narrates his attempts to respond to the riddles behind a massacre perpetrated by the distinguished Dr. David Greenwood, who had inexplicably jolted the tranquil complex only a month earlier. Moving from one enigmatic encounter to the next, Paul tries to put the pieces of the puzzle together, wondering what it is about this sterile environment that would lead a man noted for being a humanitarian activist to go on such a murderous rampage. As Paul notes, "It surprised me that he had seen enough of his colleagues to dislike them, let alone set about killing them" (39). Paul's young wife, Jane—more explicitly flagged by the book as an Alice figure—has the unenviable job of replacing the murderer: living in his house and using his office[1]—or rather, following the logic of Eden-Olympia, living in his office and using his house, since, in this place, the human body is "an obedient coolie, to be fed and hosed down, and given just enough sexual freedom to sedate itself" (17).

Dr. Greenwood's massacre is ultimately understood to be one of the negative effects of living in one of Marc Augé's "non-places," a spatial symptom of what this theorist calls supermodernity.[2] For in Eden-Olympia, "the vanguard of a new world-aristocracy" (115) begin to simply "float free of themselves" (116). Since the inhabitants spend most of their time hunched over gene-splicing equipment or stock market figures,

> there are no energies to spare for anger, jealousy, racial prejudice and the more mature reflections that follow. There are none of the social tensions that force us to recognize other people's strengths and weaknesses, our obligations to them or feelings of dependence. At Eden-Olympia there's no interplay of any kind, none of the emotional trade-offs that give us our sense of who we are. (255)

This asocial automatism has its seductive side, and Paul's wife soon joins her colleagues by "keying in the emotions she would feel that day, the memories to be cued . . . the whole programme laced with sardonic asides" (89).

Paul becomes increasingly concerned about the effect the business complex is having on Jane and attempts to resist the sinister effects of this particular "community of those who have no community." He watches himself watching a nurse attending to his injured knee, acknowledging "that I had the strong sense that we were friends who had known each other for years. Yet I had forgotten her face within seconds of leaving her" (39). Similarly, the "inept femme fatale" Frances Baring confesses, "That's the trouble with Eden-Olympia—you can't remember if you once had sex with someone" (114). Living in glass houses does not breed the paranoia that we might expect, but a sense of anonymity, even invisibility. So, when Paul complains to one of the security men, "There's no civic sense here," he is gently contradicted, with a gesture toward a nearby surveillance camera. "Think of it as a new kind of togetherness" (184)—the kind that requires "new vices" in order to both function and survive:

> The top-drawer professionals no longer needed to devote a moment's thought to each other, and had dispensed with the checks and balances of community life. There were no town councils or magistrates' courts, no citizen's advice bureaux. Civility and polity were designed into Eden-Olympia, in the same way that mathematics, aesthetics and an entire geopolitical world-view were designed into the Parthenon and the Boeing 747. (38)[3]

Things get curiouser and curiouser, however, when Paul witnesses several violent attacks on certain "foreigners"—Senegalese trinket salesmen, Russian prostitutes, Japanese photographers—by the executives of Eden-Olympia, his own neighbors and his wife's colleagues and friends.

Beneath a large Victorian mirror leaning against the wall of his living room ("It's just possible that the young Alice Liddell stared into it"), Penrose explains his radical psychosocial therapy program to the ever-inquisitive Paul, which forms the rationale behind these attacks: "We're breeding a new race of deracinated people, internal exiles without human ties but with enormous power" (256), and whether we like it or not, "the notion of the community as a voluntary association of enlightened citizens has died for ever. We realize how suffocatingly humane we've become, dedicated to moderation and the middle way. The suburbanization of the soul has overrun our planet like the plague" (263).

But in a neat twist to Freud's discontented civilization, in which the seething id is unhappily restrained by the superego's social contract, this new race of externalized beings have no internal psyche from which violence or repression is suddenly unleashed. Their subjectivity is so saturated by the ambient demands of capital that they no longer have the kind of "subjectivity" on which Freud based his theories. In order to inoculate against this particular plague, one needs to approach the psyche differently, simply because it is a different genre of subjectivity. "Any perverse elements in their lives," which are ipso facto needed to counter the numbing effects of hyper-alienation, "have to be applied externally, like a vitamin shot or an antibiotic" (259).

In his insightful reading of *Super-Cannes*, Steven Shaviro elaborates on the novelty of this "carefully metered measure of psychopathy." The point of the therapy

> is not to release deeply repressed violent and sexual impulses, but exactly the reverse: to implant those impulses into minds that previously lacked them.[4] As Deleuze and Guattari like to say, the unconscious is not given in advance; it is something that needs to be actively produced. Rather than being an outpouring of the id, the *ratissages* are deliberately staged exercises: sort of a New Age program of self-improvement, updating the old Protestant work ethic. And the therapy works. Once the *ratissages* started at Eden-Olympia, we are told, the therapeutic "benefits were astounding": the executives felt healthy, and worked harder than ever, and "corporate profits and equity levels began to climb again." In the new network economy, such controlled outbursts of ritual violence are the motor of personal growth and corporate innovation alike. By going on racist rampages, the inhabitants of Eden-Olympia are able to tap new sources of creativity; they re-establish contact with the outside world from which they have otherwise totally separated themselves. (2003, 137–38)

Such attacks represent a necessary and carefully monitored evil, argues Penrose, in an age where "[p]eople find all the togetherness they need in the airport boarding lounge and the department-store lift" (2001, 263).

The necessity for the smooth circulation of capital, along with the virtual generation of profit, annuls any old-fashioned ethics in relation to the other. "Once you dispense with morality the important decisions become a matter of aesthetics" (255), in a fusion of "might-makes-right" Nietzscheanism and "violence for violence's sake" dandyism.[5]

For in a world where the exception has become the rule (Agamben), contemporary codings of whateverbeing have their dark side, inscribing the subject as a holographic watermark within the common currency of pornography and publicity. The power elites of Ballard's world—of *our* world—have no moral center, simply because they have no center. They are people without content, executives without qualities, mere nodes in the network, but significant nodes nonetheless. (This is not to imply the salvaging of such a center but to lobby for a rigorous and sustained analysis of the opportunities provided by its absence.)

Eden-Olympia is home to the ugly face of "detraditionalization," a contemporary process whereby tradition "vanishes in the self-same discourse which purports to make its presence tangible" (Bauman 1996, 49). ("Ugly" because of its impassivity, holding the same nonexpression when listening to a board meeting as when video-recording a savage beating—"None of them look like they're having any fun" [247]—in contrast to the rather quaint ultra-violence of Burgess's [1962] and Kubrick's *A Clockwork Orange* [1971].) We do well, then, to remember that this impassive expression does *not* act as a mask concealing the horrors of humanity beneath (as we shall explore further in the next chapter), for it is a horror worn on the sleeve and on the skin.

The Looking-Glass Stage

> How would you like to live in Looking-glass House, Kitty? I wonder if they'd give you milk in there? Perhaps Looking-glass milk isn't good to drink.
>
> —ALICE LIDDELL IN LEWIS CARROLL, *Through the Looking-Glass*

Even Paul's wife begins to take part in these *ratissages*—if not directly, then in the drug orgies which follow the viewings of the videotapes. "Jane was growing up," writes Paul, "like the Alice of *Through the Looking-Glass*, and I sensed something of Carroll's regret when he realized that his little heroine was turning into a young woman and would soon be leaving him" (2001, 81). Gradually piecing together the mystery, Paul realizes that the once-idealistic pediatrician David Greenwood became so corrupted by Eden-Olympia that he found himself, almost inexplicably, involved in a pedophile ring, exploiting local refugee children and orphans. Simultaneously ironic and appropriate, Greenwood's sordid activities were operated under the cover of the "Lewis Carroll Society," while the thirty copies of *Alice in Wonderland* and *Through the Looking-Glass* in fact represented and advertised underage prostitutes for sale to the highest corporate bidders: "Here in the Rue Valentin the Red Queen was a brothel-keeper and the only looking-glasses were the smudged mirrors in the whores' compacts" (158).[6]

Through such consistent references back to Carroll's seminal books, Ballard's novel teases out the emerging logic of a new kind of meta-alienated "sense-event" (Deleuze), one which updates the now-standard, although often simplified, Lacanian mirror stage. For in the Lacanian system, the mirror stage is the moment where the infantile subject (mis)recognizes his or her image and awkwardly appropriates it as a sign of the physically and psychologically coherent self—to be worn forever like a rented suit, never quite matching the demands of the everyday roles imposed upon him or her, from within and without.

In Carroll's sequel Alice herself refuses this narcissistic splitting and chooses to wander *through* the looking glass in an experience supplementary to the first Wonderland, not even pausing to contemplate her own image. (It is no accident that the more "mature" Alice enters the liminal land through a mirror, since such a device would not, could not, precede something as "natural" as a rabbit hole.) As a result, Alice's battle with her self-perception and identity is not conducted through the scopic subject-object dialectic of the mirror stage but via the phantasmagoric dialogic of the looking glass stage.[7]

In Eden-Olympia, Jane is trapped between these two possibilities, and she soon becomes exhausted with her depraved environment and her own complicity with the system. "Too many mirrors in this house," she informs her husband, "tell me how you escape inside them" (301). In contrast to Alice, Jane lacks the willpower, imagination, or authorial support simply to turn the mirrors into shimmering doorways leading to another situation. Jane needs stronger substances to send her to dreamland: tranquilizers and narcotics, which, the narrator notes, may as well come in vials marked with "inject me" written across the labels in bold letters (323).

Indeed, Paul makes several references to his wife as "an exhausted Alice who had lost her way in the mirror world" (386), enduring a latent looking glass stage in which the coherence of her identity begins to unmoor itself from her increasingly fragile body. This Lacanian "echo"—in both the sonic sense and the Greek mythical sense—begins to untangle the knot of Jane's sense of her own discrete separation from an Other. Lacan puts it thus:

> The mirror stage is a drama whose inner dynamic moves rapidly from *insufficiency* to *anticipation*—and which, for the subject caught in the snares of spatial identification, fashions the series of fantasies that runs from an image of a *fragmented body* to what we may call the *orthopedic vision of its totality*[8]—and to the armour, donned at last, of an alienating identity, whose rigid structure will shape all the subject's future mental development. (1977, 4)

In Eden-Olympia what Lacan calls the Imaginary haunts those stolen moments of nonwork, hovering over "the swimming pools and manicured lawns" like "a dream of violence" (75). Here, as everywhere, faces act as mirrors, gently but firmly buffering the self into accepted modes of social behavior and presentation. But on the brave new Côte d'Azur, the face of the racialized Other functions less like a mirror and more like a looking glass. And as Alice herself observes in relation to her own novel surroundings, "there'll be no one here to scold me" (Carroll 1992, 112).[9]

Thus, it is worth considering whether this provisional "looking glass stage" is Lacan's mirror stage in reverse, meaning the right way round (according to the logic of double negative, since the mirror stage is *already* in reverse). If so, then the inhabitants of Eden-Olympia actually *recognize* their previous and constitutive *misrecognition.* Rather than "freeing" them from a certain ontological fracture, it in fact frees them from the "burden" of ethical conduct. Steered by the aristocratic sadism of Wilder, the scopic economy of the business center thus becomes intensely eroticized.

For instance, Paul's mistress gestures toward the mirror on the bedroom ceiling, saying:

> "If the mirror bothers you I can turn off the light."
>
> "Leave it on—I've got two of you to look at."
>
> "David liked doing that. Which was the real me, he used to ask . . . philosophy in the boudoir. The mirror was his idea." (Ballard 2001,229)

"Love" is thus figured as the hollow vertigo of two polished surfaces, facing each other, *sans subjet*—a carnival trick of smoke and mirrors.

It is with such mutations of a discourse once founded on the solid (though deluded) ego that Ballard sketches the contours of a postpsychoanalytic subject. Just like Vaughan in *Crash* (Ballard 1973),[10] Penrose is obsessed with the instrumental application of a psychopathy that *comes from the outside,* stemming from the environment itself—violence as the spasm of a particular assemblage, of which the human is only a factor, or symptom: "In many ways it seems only too apt that my guide to this 'intelligent' city in the hills above Cannes should have been a specialist in mental disorders" (3).

Such reversals potentially lead to a kind of "transitivism," whereby "the child who strikes another says that he or she has been struck; when one child is punished the other also cries" (Sarup 83). No one can be certain where he or she ends and where the other begins, except perhaps when defined and delivered through pain (or in Greenwood's case, murder). It is a Tweedledum-and-Tweedledee syndrome, only multiplied according to the exact population of the business park and the cultural moat of Cannes.

What Paul and Jane are groping toward, no less than Ballard, is an understanding of whateverbeing in its most bare and abject form, completely tangled up in the sticky webs of the spectacular market injunction. Eden-Olympia houses an entire payroll of Patrick Batemans—that is, a corporate tong of global psychos, in which colleagues can barely recognize each other because their characteristics have become illegible and interchangeable. In Brett Easton Ellis's novel *American Psycho* (1991), such anonymous individuals regularly confuse those whom they dined with last night with someone they thought they saw at a business lunch last week, or vice versa. In Ballard's novel, though, the same logic leads to an apathetic solipsism, as if the fundamental anonymity of the place creates a sense of invisibility.

Paul himself believes that "by watching our wives have sex with strangers, we dismantled the mystery of exclusive love, and dispelled the last illusion that each of us was anything but alone" (Ballard 2001, 322). But to *feel* so completely alone, wandering loose in a corporate wonderland, is not necessarily to *be* so completely alone. In fact, the "phantom community" (Durham) of Eden-Olympia marks a limit point in the atomization of modern individuals: a point where alienation is no longer simply the negation of authentic community, where technology is no longer the perversion of nature, and where love is no longer the discourse which distracts people from challenging these conceptual assumptions as false antagonists or alternatives.[11]

Like pornography, like mediation, and like other technologies which are powered by the manic dynamic between simulation and stimulation, the logic of the looking glass shatters our images of ourselves as clearly framed and delineated. It is the "danger that saves," or at least has the potential to do so, as long as we assimilate the lessons it provokes: the end of familial connections by blood, of sovereign connections by nation, and even the biopolitical connection of gender.

Unmistakably, the *ratissages* are the undesirable effects of an ontological insight only *after* it has been hijacked by the throwback resentments of psychology: "It occurred to me that psychiatry might be the last refuge of the bully" (258). Unfolding this insight in a *non*destructive direction is the task at hand, which is why Steven Shaviro felt moved enough to call *Super-Cannes* "the first great work of social theory of the 21st century" (2001).

Postcards from the Periphery

> People outside Eden-Olympia. In some way a dimension is missing. There's a lack of self-corroboration. They stroll along the Croisette, talk about their flights to Düsseldorf and Cleveland, but it's all unreal. If you stand back for a moment, tourists are a very odd phenomenon. Millions of people crossing the world to wander around unfamiliar cities. Tourism must be the last surviving relic of the great Bronze Age migrations.
>
> —WILDER PENROSE IN J. G. BALLARD, *Super-Cannes*

For Eden-Olympia's resident psychiatrist, the people milling about the Cannes waterfront outside the gates of the hermetically sealed business park have an air of unreality about them. The tourists in particular represent an incessant echo of much earlier human diasporas (just as for Paul they resemble "a huge film crew without a script" [Ballard 2001, 108]). For the sociologist Zygmunt Bauman, however, tourists signify the apotheosis of the postmodern condition, embodying the transitory law of all human commerce in an age where "[e]verything seems to conspire . . . against distant goals, life-long projects, lasting commitments, eternal alliances, immutable identities" (2003, 51).

Bauman believes that the logic of tourism has infected the globe, whereby "biographical time" restructures unfamiliar spaces and places, scrambling the geohistorical coordinates of any given location according to the ego, the camera, and the wallet, resulting in "an experience of the utmost pliability of space" (53).[12] For Bauman, tourists "pay for their freedom . . . [for] the right to spin a web of meanings all of their own" (53), mistaking and/or substituting physical proximity for *moral* proximity. This mistake is itself, he argues, an effect of an epoch in which "it is now all too easy to choose identity, but no longer possible to hold it" (50).

Within the walls of Eden-Olympia, people from all corners of the planet are also flung together in a contingent community which has more to do with a physical proximity than any ethical kind. That plural state we shall have occasion to call the "being-with" no longer feels any compulsion to answer the Heideggerean call to "the heedfulness" or even the "indifferent care" of social circuitry. Paul and Jane first encounter Eden-Olympia through the brochure, and in contrast to the all-too-common tourist mantra "This isn't anything like the brochure," the business park's newest arrivals find that it is *exactly* like living in the brochure.

Despite Penrose's insistence that there is some element lacking in the Riviera tourists—an element retained, or perhaps even acquired, by the executives of Super-Cannes—the law of touristic indifference spreads to the highest echelons of the corporate culture. This is why Alain Delages and his cronies can justify inflicting the *ratissages* on the local inhabitants without any misgivings. Indeed, in a sentence worthy of Ballard, Bauman notes that "[f]reedom from moral duty has been paid for in advance; the adventure-tour kit holds the preventive medicine against pangs of conscience, neatly packed next to the pills preventing air sickness" (54). As such, Eden-Olympians are tourists in France on a working visa, expatriates who realize that tourism "is no longer something one practises on holiday" (55).

Put simply then, Bauman believes that "The tourist is bad news for morality" (54). Much like his more disheveled cousin the vagabond, the tourist "structures the space he happens to pass through, only to dismantle the structure again as he leaves" (53). Being-in-the-world thus begins to function like a tent which one can pitch and pack up at will, and all encounters remain local, temporary, and episodic. As such, the tourist is essentially "extraterritorial," the twist being that he or she lives this permanently liminal status as a privilege, "as a licence to re-structure the world to fit his wishes" (53). Clearly, the "citizens" of Super-Cannes, who themselves live in an extended bureaucratic and psychic state of liminality, succumb to this particular culturally modified species of the will to power.

Bauman contextualizes the touristification of everyday life through terms which bring the discussion firmly back into the gravitational pull of whateverbeing. For he is responding to the current situation in which freedom—specifically, the freedom to choose our own identities—"rebounds as contingency." The ongoing effect of this is that identity itself "can be revoked at short notice or without notice—and so binds no-one, including the chooser" (51). Bauman's argument seeks to reestablish a moral code in which we no longer throw out the baby (i.e., morality itself) with the bathwater (i.e., universal foundations for morality).

For him this admittedly hazardous task begins and ends with "the moral capacity of the self" and a "voyage of self-discovery" (58). Presumably, such a voyage is not undertaken as a package tour but as a pilgrimage, or as some other nontouristic mode of personal movement and experience—one which navigates away from the "relentless pulverization of collective solidarity" (57) and toward "the wondrous aptitudes of sociation, rather than the coercive resourcefulness of socialization" (58).

However, we need not be as quick as Bauman to see warning lights accompanying every situation wherein people engage in nothing "but the briefest and most perfunctory of encounters" (as we shall see later, particularly in the case of Wong Kar-wai). Such fleeting, asymptotic meetings have the potential to force "us" to see the constitutive anonymity and collectivity of (social) existence. Where the multitude shudder at the thought of being considered part of the multitude ("Put that map/camera/Hawaiian shirt away! We look like tourists!"), incessant global tourism reminds us of the violent banality of deracination.[13] The (im)posture of the sensitive individual—the *true* traveler, in sharp relief from the homogenous tourist hordes—is a myth nurtured by practically *every one* of these people, who in fact collectively *make up* the hypothetical mass of the selfsame tourist hordes. (A constitutive paradox, symbolized by tourist guidebooks which claim to guide the reader away from tourists.) The stereotyped "tourist" thus functions as Other, not only for the sociologist but also the tourist him- or herself.

"No-one but the tourist is so blatantly, conspicuously dissolved in numbers, interchangeable, depersonalized" (54), writes Bauman. And yet, according to his own logic—in which "we" (Westerners, at least) have all become tourists—we are *all* undergoing this dissolution of identity. Like the snap-happy tourists standing in front of Don DeLillo's "most photographed barn in America," we no longer see the barn. Moreover, we no longer see each other, and "we can't get outside the aura. We're part of the aura" (1985, 12–13).

On the surface, Eden-Olympia seems to bear out this claim by Bauman: "Ethnic herding and confessional flocking together take over when the collective responsibility of the *polis* fizzles out. The dissipation of the social rebounds in the consolidation of the tribal" (1996, 57).[14] But as we have seen, this is not a *return* to tribalism (in the Dionysian mode celebrated by Michel Maffesoli), but rather an attempt to consolidate a new, essenceless tribal form, in which belonging is purchased by the price of silent complicity and lasts only as long as your contract of employment.

Bauman believes that moral responsibility occurs only in situations where there is a profound recognition of "the uniqueness—irreplaceability—of the moral subject" (54). However, such a belief fails to acknowledge the extent to which these same concepts work to camouflage the crucial moment when "inimitability" actually *authorizes* violence and negation of all kinds, a violence which "violates what it exposes"—that is, being singular plural.[15] The question of essence once again becomes paramount, since the notion of the morally capable "unique individual" often smuggles severely compromised definitions into the discussion, against its own interests. Indeed, such smuggling is conducted in the manner of a suicide bomber and with analogous motives—that is, to wreak havoc on those perceived to possess a different kind of essence. No matter how much it insults our delicate sensibilities, the fact remains that we *are* eminently replaceable on the macro or social plane. (To the daily reminder of carnage and death, the world shrugs and says, "Life goes on.") And that is why we must refuse to draw lines, distinctions, borders, and conclusions prematurely. Precisely *because* everyone is a potential lover, he or she also is a potential victim of our hatred, and we foreclose on our own futures when we internalize an identity-as-essence or when we count off our conditions of belonging like rosaries or worry beads.

Jean-Luc Nancy captures this founding tension between our singularity and our inbuilt obsolescence when he writes:

> There is a common measure, which is not some one unique standard applied to everyone and everything. It is the commensurability of incommensurable singularities, the equality of all the *origins-of-the-world*, which, as origins, are strictly unexchangeable [*insubstituable*]. In this sense, they are perfectly unequal, but they are unexchangeable only insofar as they are equally with one another. Such is the sort of measurement that it is left up to us to take. (2000, 75)

Thus, according to the techtonic movements of twenty-first-century globalization, the periphery folds back into the center, perhaps even generating the illusion of a center in a centerless network. For Baudrillard and Bauman, tourists send postcards from imaginary realms like Rome, Bali, and Cairo to equally chimerical destinations such as Vermont, The Hague, and the *Freedom Ship*.[16] Even U.N. soldiers and other so-called peacekeepers take photographs and buy trinkets for the "folks back home"—trophies of non-experiences in non-places.[17] Movement, moreover, does not necessarily denote "freedom," as the famous case of carp illustrates: A carp will literally die of boredom in a large, empty pond but will swim around happily and indefinitely if a rock is placed in the center—its "happiness" thanks to, presumably, the illusion of options.

The German philosopher Peter Sloterdijk has considered the implications of such a phenomenon when applied to human beings, concluding that "the keeping of men in parks or stadiums seems from now on a zoo-political task" (2001, n.p.). Within this schema we must of course include business parks, such as Eden-Olympia: "I knew that he [Penrose] saw me as another of his experimental animals," writes the narrator, "to be stroked through the bars as I was fattened for yet another maze" (Ballard 2001, 279). As much as for Wilder Penrose as for Plato,

> What is presented as reflections on politics are actually foundational reflections on rules for the maintenance of the human zoo. If there is one virtue of human beings which deserves to be spoken about in a philosophical way, it is above all this: that people are not forced into political theme-parks, but rather put themselves there. Humans are self-fencing, self-shepherding creatures. Wherever they live, they create parks around themselves. In city parks, national parks, provincial or state parks, eco-parks—everywhere people must create for themselves rules according to which their comportment is to be governed. (Sloterdijk 2001, n.p.)

Even corporate think tanks themselves, such as the Union of International Associates, openly discuss the irony that "knowledge workers, unknowingly entrapped in such environments, might well be described as being in a *concentration camp*. It would be their very concentration and focus that would prevent them from experiencing how they were constrained" (Judge, 2001), a comment which could easily have served as the germ of Ballard's book.

Sloterdijk, moreover, writes, "Man, who is confronted with a library, becomes a humanist. Man, who is confronted with a computer however, becomes someone for whom we have no name yet. Here a post-literary, post-humanist type of man is developed" (2001). In *Super-Cannes* this post-humanist type of man is not so much the Nietzschean *Übermensch* but the infinitely degraded corporate version. Despite being compromised by (what can only be presumed is) its ironic subject ("Man") as well as the ambiguous tone of the verb "confronted" (as if libraries and computers simply fall out of the sky, like the monolith in *2001: A Space Odyssey*), Sloterdijk's formula is a useful reappraisal of the recent shifts in discourse network. The knowledge workers in Eden-Olympia crunch binary, genetic, mathematical, and economic codes. They inhabit the avant-garde of instrumentalist technicity, dealing as much in pictorial symbolic iconography as in the linear alphabetization which spawned their scientific forefathers. As such, they do indeed represent an emerging posthumanistic, if not exactly posthuman, type (in the terms so eloquently presented in the work of N. Kathryn Hayles [1999]).

For Ballard the two nominal halves of the business park, Eden and Olympia, have invoked the techniques of Hephaestus to fuse themselves into a neoclassical unity: a pantheon of demigods untainted as yet by the moral injunction of Original Sin (also known as "conscience"). Ultimately, this feat of socioeconomic and autogenetic engineering means that "The people here have gone beyond God. Way beyond. God had to rest on the seventh day" (Ballard 2001, 202).

FOUR

Facing the Interface

> The face is exposed, menaced, as if inviting us to an act of violence. At the same time, the face is what forbids us to kill.
>
> —EMMANUEL LEVINAS, *Ethics and Infinity*

I am sitting on a stationary bus, leaning my head against the window and feeling pensive; as people often do when they are alone on public transport. Presently, another bus pulls up right beside me, forcing a stranger's face directly into my field of vision, only three feet away and in a similar pensive, head-leaning position. She is rather indistinct, but her presence is undeniable, and there follows an extended moment of awkwardness as we wait for the lights to change. We exchange glances but can't very well continue staring into each other's eyes, at least not without doing some kind of existential violence to our psychic equilibrium. For "staring people out" is either the hostile game of teenagers, misogynists, and racists or the deliciously risky game of lovers. And as random commuters, we are neither of these.

Since two heavily vibrating panes of glass separate us, we cannot smother the discomfort of the situation through small talk. So, we both stare down at the road, held to ransom by Levinas's originary "shame of being": "Do I have the right to be?" (1985, 121). Usually, phatic speech covers these ontological doubts with the banality of everyday encounters, suspended in the sticky amber of reification: "How are you?" to the neighbor. "Have a nice day" to the cashier. "How about that local sports team?" to the boss. Hence the panic which can seize people displaced outside their linguistic boundaries; who have to *face* the other unclothed by language, existentially exposed to a rudimentary confrontation with alterity. "It is the most naked," says Levinas, speaking of the face, "though with a decent nudity" (86).[1]

The epiphany of the face: an event which lies dormant in every meeting yet sometimes blossoms beyond the humdrum of quotidian contemplation. The structure of such an intense encounter is captured in the otherwise forgettable Baz Luhrmann version of *Romeo + Juliet* (1996), where an initial glimpse of the future beloved (through the fish tank, in this case) provokes the sudden loss of self in the face of the other. It is both a revelation and a rediscovery of Narcissus (who, as Agamben shows in his book *Stanzas*, was not "narcissistic"). The lovers shed the awkwardness of being for a luminescent curiosity for the other. As such, "love at first sight" is a triumph over perceived ontological shame—or at least a profound acceptance of it, which amounts to the same thing. "The true union or true togetherness is not a togetherness of synthesis," states Levinas (the resident authority on such matters), "but a togetherness of face to face" (77).

Levinas sees the face as both an order and an ordination for the subject, as an essential etiquette taken to the highest level of ethics, "an original 'After you, sir!'" (89, 97).[2] More importantly for Levinas, face and discourse are tied, since it is the face which speaks: "[T]he *saying* is the fact that before the face I do not simply remain there contemplating it, I respond to it" (88). But as we have just seen—through both the shuddering bus windows and the Capulets' fish tank—there are revealing moments when speech is denied, impossible, inadequate, or simply unnecessary. In the case of love, we search this radiant face for the quivering presence of the code itself, before content and before Proustian signs compel us to interpret, deduce, and decode the beloved through the grids of jealousy, egotism, and the narcissistic feedback loop of literary love. Hence Levinas's recognition of the fact that "across all literature the human face speaks—or stammers, or gives itself a countenance, or struggles with its caricature" (117).[3]

Staring into the face of the other calls for an untangling of this complex interplay between vision, speech, discourse, and seduction. For it is not simply a case of "drinking" the other in through the eyes, since there are resistances, "black holes," blockages, projections, and introjections in the libidinal-scopic event. This is why Levinas believes that "[t]he best way of encountering the Other is not even to notice the color of his eyes!" (85). Levinas thus refuses vision as the medium for comprehending "the authentic relationship with the Other," whether it be figured as a social response or as a more profound *a priori* responsibility (87–88), since vision "is a search for adequation; it is what par excellence absorbs being" (87), and an authentic relationship renounces all attempts to appropriate the other. As such, for Levinas at least, the face signifies signification itself: "signification without context . . . [since] the face is meaning all by itself" (86).

Deleuze and Guattari, however, approaches the same fundamental configuration from a different perspective, since for them the face is not a given to which the subject responds but is actively produced through eye-to-eye contact. The face is less the catalyst for an epiphany than a by-product of the "four-eye machine" (1999, 169). It is therefore illuminating to contrast Levinas's reading of the face with Deleuze and Guattari's "defacialization," in order to trace the shift from intersubjectivity to a nondialogical understanding.

For Deleuze and Guattari (henceforth simply the authorial assemblage "Deleuze," for convenience), the face is "a very special mechanism" situated at the intersection between the strata of signifiance and subjectification. "Signifiance is never without a white wall upon which it inscribes its signs and redundancies. Subjectification is never without a black hole in which it lodges its consciousness, passion, and redundancies" (167). The face is thus the result of a "white wall/black hole system" inherent in Western modes of perception and inscription, whereby the semiotic gravitational pull between "surface" and "hole" (also witnessed in the techniques of landscape design) is most apparent.[4]

In contrast to Levinas, in which the face signifies the Infinite, Deleuze sees the face as simultaneously a "a monstrous hood," "a multiplicity," and most certainly "a politics." The process dubbed by Deleuze as *visagéité* (faciality) is thus an operation of "the abstract machine that produces faces according to the changeable combinations of its cogwheels" (168). In other words, the face is a *technology*, and one not too far removed from Stiegler's handshake.[5] For Deleuze the face is more or less frozen into place by the territorializing forces of subject formation and sense making, and as such is fundamentally different from the head and the body, both of which tend to resist the tyranny of the signifier. And yet this resistance can buckle under the relentless pressure of facialization: "when the mouth and nose, but first the eyes, become a holey surface, all the other volumes and cavities of the body follow" (170).[6] Thus the entire body can become facialized.

Even more important than the overcoding of the body-head by the face is the relationship of the face to individuality and humanity in general. *Contra* Levinas, Deleuze insists that "there is even something absolutely inhuman about the face," since even the most minor distortion, or switch in perspective, can collapse the reassuringly fundamental distinction between face, mask, and landscape. The grotesquerie and accompanying spectator fear of clowns has only increased in direct proportion to the value we place on being able to read individuality and intention into the face. Likewise, the grimaces and contortions of the orgasmic face are either neutralized in the anonymous crevices of pillow and shoulder or endured and forgiven by love, boredom, or indifference. (The male face, wracked by orgasm on camera, is *always* ludicrous, since it grotesquely documents the dissolution of masculinist mastery. The simulated orgasms of celluloid women, though, represent the *jouissance* of Womankind and as such have no individuality or poise to lose, at least according to the patriarchal-scopic mode which seeks out and documents such moments.)

Deleuze's position on the traditionally assumed affinity between self and face can be gleaned from the following statements: "The face is not an envelope exterior to the person who speaks, thinks, or feels" (167). "Faces are not basically individual; they define zones of frequency or probability" (168). "It is not the individuality of the face that counts but the efficacy of the ciphering it makes possible, and in what cases it makes it possible" (175). "You don't so much have a face as slide into one," and "faces . . . choose their subjects" (180), not the other way around. We have been well prepared for such "asignifying and asubjective" interpretations by the reading of *Go-Sees* in this book's introduction, whereby sheer plurality creates the conditions for potential whateverbeing to emerge.

For Deleuze, then, the face is not universal, because it is the machinic unit of measure invented by the "White Man." Its purpose is to bring all other modes of being into the facialized, homogenized human economy (cultural others, certainly, but also pets, clouds, shrouds, and other creatures and objects which are vulnerable to the transcoding of the abstract machine).[7] Take, specifically, the anthrocentric codings of animals: dogs "look" cute, cats "seem" snobby or narcissistic, and sharks "appear" cruel. Through such daily projections, the face reveals itself as the original screen of Being (in the sense of "screening" calls).[8]

But while one function of this machine is to bring everything into the anthropological sphere, the other is to then measure, scan, record, and organize it under the regime of faciality. Thus, while sitting at a sidewalk café "scoping" the street for attractive and/or familiar faces, we enact the logic of police profile-matching software programs, which match surveillance footage with archived criminal identikits. In allowing our friends and family through the door but excluding strangers, we mimic the commands of the recognition software used to facilitate passenger processing in airports.[9] However, according to Deleuze, the chances are that it is already too late, that "you've been recognized, the abstract machine has you inscribed in its overall grid." And so, "at every moment, the machine rejects faces that do not conform, or seem suspicious." What the media spectacle values above all else ("Rita Hayworth gave good face"—Madonna, "Vogue" [1990]), Deleuze dismisses as "the black hole of subjectivity as consciousness or passion, the camera, the third eye" (Delueze and Guattari 1999, 168).

Before we consider further the complicities of the media, information technology, and the "redundancy of resonance and coupling" (i.e., "romantic love"), we must consider some recent scientific discoveries concerning vision and the face, as mediated by the abstract machine of the brain.

The Nastiest One of All

> In most cases . . . memories supplant our actual perceptions, of which we then retain only a few hints, thus using them merely as "signs" that recall to us former images. The convenience and the rapidity of perception are bought at this price; but hence also springs every kind of illusion.
>
> —HENRI BERGSON, *Matter and Memory*

To grasp the face's truth means to grasp not the *resemblance* but rather the *simultaneity* of the visages, that is, the restless power that keeps them together and constitutes their being-in-common.

—GIORGIO AGAMBEN, *Means without End*

For several decades scientists have known that specific parts of the brain are used to reconstitute and recognize the visual data that make up our phenomenological world. However, it has not been until very recently that researchers have realized that different neurological networks are used to recognize faces in distinction from other "nonfacialized" objects. More than twenty different neurological systems make up our "wetware's visual recognition system," but only one of them, it seems, is used for the face. We know this thanks in large part to two diametrically opposed brain damage cases featured in the BBC documentary *Brain Story: The Mind's Eye.*[10] One man can recognize faces but not objects, and the other can recognize objects but not faces. In the latter case, more relevant to us in the present context, the subject is always *cognizing* faces but never *re*-cognizing them. This dilemma becomes clear in his conversation with a research scientist who is showing him a series of slides for the purposes of identification:

Researcher: "So, what's this?"
Subject: "That's an apple."
Researcher: "Okay, next?"
Subject: "That's a plate setting with a knife, spoon, fork, and a plate."
Researcher: "And who is this?"
Subject: [Long pause] "I don't know. I don't know."
Researcher: "Young woman? Old woman?"
Subject: "To be honest, I'm even having trouble answering that question."
Researcher: "Let me give you some biographical details: She's a movie star from the 1950s, very glamorous, she sang 'Happy Birthday' to President Kennedy once."
Subject: [Pause] "Marilyn Monroe?"

Later in the program this same man elaborates on his orientation toward the banality of everyday faces:

I'm asked . . . sometimes if people just look all the same, and my answer to that is "No, they don't all look the same"; and people say, "Well, why don't you recognize them?" and the answer is "They don't all the look the same, but they don't all look like John Smith." None of them look like anyone: They all are faces, but they are not recognizable as *anyone's* face. Any of them.[11]

As the researcher then explains, "He can see the texture of the skin, he can see the individual features fairly well, but what he doesn't see is the *totality* of the face." Moreover, this subject's affliction applies just as much regarding himself as others:

Researcher: [Changing slides again] "Okay, who's that person?"
Subject: "Hmmmm. Don't know."
Researcher: "Any sense of familiarity?"
Subject: "No."
Researcher: "What about the child's face, looking over his shoulder? Any familiarity there?"
Subject: (Thoughtful pause) "This is the nastiest one of all, isn't it? This is me."
Researcher: "Yes. How do you feel about looking at a picture of your own face and not recognizing it?"
Subject: "For me, it is *a* face. It's not *my* face. It's *a* face. And there is some sense of incompleteness there. But, uh . . . [long pause] . . . so be it."

It is in this rather tense pause, between the subject's acknowledgement of "incompleteness" and his ontological resignation ("so be it"), that we rejoin the slippery bond between whateverbeing and the face. The latter is traditionally taken to be a "stamp of personality," which imprints itself in our waxy Cartesian substance sometime after puberty, when our features settle into, or rather reflect, our personality (as we saw in Proust's depiction of the underformed features of the girls' faces on Balbec's beach). In social life we thus rely on using the face as an index of identity (albeit one easily disturbed and disrupted by the uncanny existence of identical twins and "chimeras").[12] The face is a text—a unified and separate set of signs, yet open and exposed, in Levinas's sense. The face invites us to attempt to decode it, unless the pre-veiling law commands certain sectors of the community to deny any such attempts at existential access (as in certain Muslim communities). That is, we can try to "read" people's faces unless they are veiled or masked or otherwise hidden from us.

Hence the significance of the mask on the pillow in Kubrick's *Eyes Wide Shut* (1999), a film whose title could well be applied to both the brain-damaged patient, who can't recognize his own wife for neurological reasons, and the film's protagonist, Dr. Harford, who can't read his own wife's domestic mask for psychosocial reasons. The reversible slippage between face and mask betrays the anxieties of knowledge about the other: How well do we *really* know our loved ones, even ourselves? This is the question behind countless tales of body swapping and identity switches—specifically, the consistency and resiliency of a coherent "self" when transplanted into a different physical context, symbolized most powerfully by a new face. Overwhelmingly, these narratives generate their own power from the hope that our identities can withstand the metaphysical disorientation of staring into the mirror and seeing an unfamiliar countenance. The violence of this misrecognition breaks the narcissistic, self-reinforcing circuit, and yet "love"—bestowed on the one who sees the subject for "who they really are"—salvages the radical doubt that comes from being simultaneously self and other.

This is certainly the moral lesson of John Woo's provocatively ludicrous film *Face/Off* (1997), in which the long-suffering Hollywood wife must come to terms with having unknowingly shared her marital bed and body with a villain surgically wearing her husband's face. Just as difficult is having to learn to love her husband again, although he "has" the face of the man who killed their young son (at least until the operation which will finally and efficiently erase all these traumas). *Face/Off*, along with its cinematic play with mirrors, is a response to the received cultural wisdom that "you get the face you deserve." The hero refuses to relinquish his thirst for revenge and is thus condemned to live with the face, and therefore partly within the identity, of his enemy. By the end of the film, however, the family is reunited through a particularly audacious narrative poaching—in this case, of a gangster moll's orphaned son, who then replaces the hole left by their murdered son. (In a related piece of restoration, the teenage daughter, who has been hiding her grief and adolescent confusion behind a mask of makeup, finally shows her true, unsullied face during the *dénouement*.)

Once again, we confront that constitutive tension between irreplaceability and interchangeability.

The face and the proper name act in cahoots via the bureaucratic interpellations which largely produce one's "identity." It has been Foucault's great contribution to demonstrate how modern discourse constantly fosters, "identifies," and regulates subjects in order to maximize the efficiency of its institutions (including, of course, capital itself). This is why something like "lifestyle consumption" is largely a matter of maintaining or creating an identity from a set of severely circumscribed and compromised options. Hence the intrinsic impotence of "identity politics," which is part and parcel of the restricted economy it allegedly opposes.[13]

Accordingly, the face shrink-wraps itself around the subject in a manner akin to the newborn alien covering the head of its unfortunate host, where it sucks nourishment from the warm human body, in Ridley Scott's classic science fiction movie. ("The face is a horror story"—Deleuze.) Under such a bureaucratic regime, surgically acquiring a new face becomes the closest thing we have to Deleuze's desired "defacialization," since it allows a certain amount of distance from our former selves (along with its associated responsibilities, histories, expectations, disappointments, and overdeterminations). And yet becoming-other under such conditions creates its own risks.

"Please Sir, I Want Some More"

> Modern intersubjectivity is today being founded on a void faciality, on a blind face-to-face between two vapid looks.
>
> —MARSHALL BLONSKY, *On Signs*

> What dreadful bondage, the bondage of my face—or one of my former faces.
>
> —JORGE LUIS BORGES, "The Aleph"

John Frankenheimer's remarkable cult film *Seconds* (1966), based on the novel by David Ely, goes further than *Face/Off* (Woo 1997) in the depiction of the social disorientation which would understandably follow radical facial surgery. The initial credit sequence begins with disturbing close-ups of those anonymous "black holes" of the face, starting with the eyes, moving to nostrils, ears, and mouth. We are soon introduced to Arthur Hamilton (played by John Randolph), a nondescript man who has recently been "sponsored" by a friend of his, a friend who had hitherto been classified as "deceased" but in fact was the beneficiary of just such a sponsorship himself several months earlier. As the puzzle unfolds, we learn of the clandestine existence of an organization called simply the Company, which convinces these men (and they are all men, it seems) to start life again with different names and different faces. This Company, founded and run by an amiable yet sinister doppelganger of Colonel Sanders, gives men of a certain age the chance to have "seconds," to sidestep the desperate automatism of middle age through a complex process of physical and social "rebirth." In return, most of each man's assets are turned over to the Company, although the widow (grieving or otherwise) is left with the house and enough to live on. Arthur soon realizes that once a man has been sponsored, he has no choice in the matter, for the Company relies on complete secrecy in order to operate.

When the bandages come off, Arthur—now "Antiochus 'Tony' Wilson"—sees a monstrous visage in the mirror, scars and stitches like Frankenstein's creation. The plastic surgeon, however, reassures him that this particular procedure was something of a milestone and he has been the beneficiary of the surgeon's "best work." As the scars heal, his features settle into the recognizable face of the young Rock Hudson—for this is indeed who is playing Mr. Wilson—and the protagonist's new lease on life is effectively sealed. (Moreover, the signature which first inks the deal is one of the best meta-edits in Hollywood cinema, cutting seamlessly from pen to scalpel.)

Wilson is relocated to California, where he poses as an artist, a pose which is supposed to become a reality over time, with practice (this profession was extracted from his unconscious wishes under hypnosis). "You will be in your own new dimension," the psychiatrist tells him, "alone in the world. Absolved of all responsibility, except your own interest. Isn't that marvellous?!" Despite his newfound good looks, beachfront bachelor house, and "ideal" lifestyle, Wilson is troubled by the unfamiliarity of his surroundings, especially with the unfamiliar face that masks Arthur Hamilton's all-too-familiar thought patterns. Alcohol seems to be the only thing which soothes the nerves—nerves that have become more frayed and exposed by this second case of *Geworfenheit* ("thrownness"), all the more alienating for happening to an established, albeit miserable, identity.

When still in his previous persona, Arthur Hamilton had admitted to the Company Director, and eventually to himself, that there was absolutely nothing to live for in his former life: "Is there anything at all?" the Director asked. "I expect to be president of the bank before too long," he replied, uncertainly. "And I have my boat in the summer. We have friends." The Director then slowly repeated the question, for effect: "Anything . . . *at all?*" After the rebirth, Wilson cannot seem to reconcile the contradiction of having his old eyes staring out from a new set of features. Indeed, his headaches are massaged away by fingers with a different set of prints.

Things begin to improve when Wilson meets Nora Marcus (played by Salome Jens), a Bohemian escapee from Middle America. She is a bit like him, although more extroverted and spontaneous. She attempts to read the enigma behind his sullen, handsome face, claiming that there is an element which "doesn't emerge pure; it pushes at the edge of something still tentative, unresolved, as if somewhere in the man there is still a key unturned." For a moment he is surprised: "That's quite an analysis." She then admits, following the logic of newspaper horoscopes and telephone psychics, "Not really. When you come to think of it, it sort of fits everybody."

It is Nora who also takes Wilson to the Bacchanalian wine-making festival in the hills behind Santa Barbara. After some futile attempts to resist the orgiastic celebrations, he follows her by jumping naked into the grape-filled vat with a throng of other people, shouting with manic glee and connecting for a moment with "the practice of joy before death." This laughter marks the symbolic rebirth, and the wine-spattered baptism, of a child of Dionysus.

Its effects begin to wane later at home, however, as the same god of viniculture forces him to confess his terrible secret and his true identity. The Company spies, including Nora herself, bring Wilson-Hamilton back to HQ for his indiscretion. Although his drunken confession was heard only by insiders and therefore didn't compromise the operation on this occasion, he has proved himself a liability and blown his chance. Wilson-Hamilton's pleas to "start again," to "do it right this time," fall on deaf ears. Indeed, the audience is as skeptical as the Company boss that he can readjust more successfully next time, not least because of the "high percentage of failures" among reborns. And yet, he is reassured that things will be all right.

At the end of the film, Wilson-Hamilton thinks he is being taken to surgery for yet another face. But in *this* Company existential orphans are only allowed seconds, not thirds. The end credits roll over Wilson-Hamilton's hideous screams upon realizing that he has been signed over for "cadaver use" in the next staged murder: He will soon be a substitute corpse for the next "sponsored" man.

The Clarity of Cloudy Vision

> Be only your face. Go to the threshold. Do not remain the subjects of your properties or faculties, do not stay beneath them: rather, go with them, in them, beyond them.
>
> —GIORGIO AGAMBEN, *Means without End*

> "I can hear her sweet country soul in every digitally encoded bit."
>
> —LISA SIMPSON in *The Simpsons*

So much for the face, but what about the interface?

In William Gibson's 1996 novel *Idoru*, the human protagonist, Laney, blushes when he finally finds himself face-to-face with the virtual pop star Rei Toei, whose bodily presence was in fact projected by a sophisticated hologram device: "He looked into her eyes. What sort of computing power did it take to create something like this, something that looked back at you?"(237). This scene is significant because it suggests that we do not have to be exposed to a fellow human in order to feel thus exposed. But then the question arises: To what are we being exposed?[14]

Laney himself is a second-generation hacker from Gibson's cyberpunk world, the spiritual nephew of Case from the original cult classic *Neuromancer* (1984). Laney's skills in surfing the web are slightly less abstract than Case's in the first cybernovel, but no less poetic, since he can defocus into a "nodal" state, encouraged by blacklisted drugs and well-funded intelligence programs. This nodal state allows him to read patterns in an intuitive way, almost subconsciously: "He had a peculiar knack with data-collection architectures, and a medically documented concentration-deficit that he could toggle, under certain conditions, into a state of pathological hyperfocus" (25).

Laney is thus capable of tracing

> the sort of signature a particular individual inadvertently created in the net as he or she went about the mundane yet endlessly multiplex business of life in a digital society. Laney's concentration-deficit, too slight to register on some scales, made

> him a natural channel-zapper, shifting from program to program, from database to database, from platform to platform, in a way that was, well, intuitive. (25)

Thus, our hero is a kind of native guide to the future internet; someone who describes his vocational experience as "seeing things in clouds. . . . Except the things you see are really there" (148).

And it is this specialized cognitive talent which provokes Laney's particularly intense experience when faced with the holographic virtual pop star. For when he looks at her, looks into her eyes: "He seemed to cross a line. In the very structure of her face, in geometries of underlying bone, lay coded histories of dynastic flight, privation, terrible migrations. . . . The eyes of the idoru, envoy of some imaginary country, met his" (175–76). The idoru's face is not flesh but a different kind of information. It is the tip of an enormous iceberg, which induces nodal vision for Laney "as narrative" (178). Moreover, these narratives, described as the virtual dreams of a virtual being, are even recorded and stored as promotional material. "We don't 'make' Rei's videos," says the idoru's agent, "not in the usual sense. They emerge directly from her ongoing experience of the world. They are her dreams, if you will" (237).[15]

The idoru even seems to flirt with Laney, obeying some kind of advanced algorithms lying dormant in the libidinal logic of Alan Turing's experiments. "You dream as well, don't you, Mr. Laney? . . . That is your talent. Yamazaki says it is like seeing faces in the clouds, except that the faces are really there. I cannot see the faces in clouds, but Kuwayama-san tells me that one day I will" (ibid.). In this case, and in direct contrast to the English expression, cloudy vision represents clarity on another level of legibility. What on the one hand could be dismissed as the habitual projections of faciality—seeing faces in the clouds—becomes rather the portrait of a reterritorializing posthumanity.[16]

Indeed, Gibson's digital-primal scene prompts a variety of questions concerning the role of what is currently called F2F contact in a world considered by many as overmediated and impersonal. It could be said that Laney and Rei Toei are locked into one of Bergson's "zones of indeterminacy," in which the status of "living images" are suddenly brought into crisis. So saying, if we can no longer make any definitive distinction between objects, subjects, and images—as Bergson initially claims in *Matter and Memory* (1988)—then how are we to distinguish between a hologram and a "real" human being (or, for that matter, between Roland Barthes's mother and a *picture* of Roland Barthes's mother)?

Such questions are especially pressing when we acknowledge that the ontological logic of the interface is simultaneously ubiquitous and neglected. For while the *Oxford English Dictionary* defines *interface* as "a means or place of interaction between two systems," few people who employ the term as a new-media cliché pause to ask the crucial question, What criteria do we use to judge an independent system in a networked age, when boundaries are far from obvious?

Deleuze's *faciality* can now be reposed as a question of *inter*-faciality, with an emphasis on the digital rather than the mechanical. Such a perspective allows us to think through the possibility that the face is *always already* an interface, one which can only exist *in relation to*, and not prior to, any possible meeting. The originary fractured overlap of so-called subjects—flagged by Heidegger's concept of *Mitsein* (with-being) and developed by Jean-Luc Nancy in his discussion of "being singular plural" (that is, being simultaneously both singular *and* plural)—subverts the common mode of egocentric, atomistic thinking on the matter of social relations.

As Paul Morris puts it:

> Beginning from the assumption of the existence and reality of the individual, who then, and only then, forges links between herself and other separate and discrete individuals, it becomes all but impossible to conceive of any sort of community at all. One plus one plus one plus one plus one just never seem to add up to more than a number! This separating out of individuals . . . does not let us put Humpty Dumpty together again. (1996, 226)

In other words, we have yet to respond adequately to the wealth of writings which encourage us to consider the self as an emanation of the many, rather than the many as the aggregate of selves. The idoru herself is a figure who can help us think being singular plural, since "Rei's only reality is the realm of ongoing serial creation. . . . Entirely process; infinitely more than the combined sum of her various selves. The platforms sink beneath her, one after another, as she grows denser and more complex" (Gibson 1997, 202).[17]

This brings us back to the significance of Laney's blushing under the uncanny gaze of Rei Toei. Agamben states that "in shame we are consigned to something from which we cannot in any way distance ourselves" (1999a, 105), that is, to ourselves. Referring to Levinas once more, Agamben defines shame as the experience of "being chained to oneself, the radical impossibility of fleeing oneself to hide oneself from oneself, the intolerable presence of the self to itself" (105–106). We can respond to this shame in different ways, including escape from the self (via, for instance, a portal into John Malkovich's brain) or wallowing masochistically in the enchainment of self (as Rousseau does when he exposes himself to washerwomen in the streets of Turin, or equally to the literati in his written confessions). Shame is thus the disorienting simultaneity of subjectification *and* desubjectification—in rather old-fashioned language, "an encounter between man and Being" (106–107).[18]

In the case of Laney, it is an encounter between a man and a digital being, a point of unprecedented emergence, which can nevertheless be placed on a continuum of encounters with the human and the nonhuman. This encounter (which occurs in a Japanese nightclub, called the Western World, decorated by "chemically frozen frescoes of piss") echoes Heidegger's and Rilke's uncertainty when faced with the existential "openness" of animals, most notably the tiger.

From this perspective, it could be claimed that Heidegger was too specific when he insisted that modernity was founded on the forgetting of Being. Perhaps earlier epochs were similarly blinded by the light of ontology and were thus also forced to drape the dressings of daily life over the harsh glare of actual existence. In other words, to truly, *profoundly* recognize the other, in some kind of Heideggerean or Levinasian sense, is both dazzling and distressing. Even the Greeks admitted that they perceive the earthly world "but through a glass dimly" (Plato 1999b, 811), and that to wrench the gaze from the flickering shadows that comprise our days is to "suffer sharp pains" and a distressing glare (Plato 1999c, 265–68). Plato's thoughts on the matter sometimes read like a metaphysical case of conjunctivitis: "[S]o does the stream of beauty, passing the eyes which are the natural doors and windows of the soul, return again to the beautiful one. . . . [H]e appears to have caught the infection of another's eye; the lover is his mirror in whom he is beholding himself, but he is not aware of this" (1999b, 817–18).

One could be tempted to wonder whether this is why we enjoy gazing into a loved one's face when she or he is asleep, unhindered by what Henry Miller called "the engulfing immediacy" of it. Metaphorically speaking, during the waking hours we are blinded by the eyes of another, as if the searching pupils were in fact headlights of the soul, permanently switched to high-beam. And this would also account for the discomfort which accompanies eyes which lock for too long, as we have already mentioned, or the dazzling effects of love at first sight, an exceptional mechanism which allows for this momentary witnessing of the other's presence in Being, as a piece of smoked glass allows the viewing of a solar eclipse.

But there is a danger here of subscribing to a universal, trans-temporal definition of shame, specifically as it is tied to the notion of saving face. For while an anthropological argument could be made that the fear of exposure or humiliation is present in some form within all cultures, the social significance of both its cause and effect vary greatly. For instance, if Clifford Geertz is to be believed (and there is no immediate reason why he shouldn't be), the typical Balinese has a mortal fear of being considered an individual, being apart from his or her social status. We are told that the Balinese word *lek* denotes a combination of "shame," "stage fright," and "fear of exposure."[19] Thus, when "a breach in the public/ceremonial persona" occurs, "the immediacy of the moment is felt with excruciating intensity and men become suddenly and unwillingly creatural, locked in mutual embarrassment, as though they had happened upon each other's nakedness" (62–64).[20] In stark contrast, my recent experiences in Holland make me doubt whether the ontological subtleties of personal shame ever cross the mind of the Dutch. Moreover, the garrulous sexlessness, and so-called "shamelessness," of very young children should also warn us against ascribing a transcendent sense of shame, distributed at birth, equally to all. Hence, shame is something we learn; and to different degrees.

This is to say that Gibson's Eastern encounter in the Western World may seem like a trivial moment in a less-than-canonical book, but for those trained to defocus, it can act as a nodal vision for understanding the fact that the very term *intersubjectivity* defines and limits the terms of debate or analysis. Just as the face is always already an interface, subjectivity does not automatically precede intersubjectivity.

It thus becomes clear that the face—or more specifically, our reading of the face—is a technology of belonging, of bringing people into the communal fold. At the same time, it is a register of those who should be *excluded* from this same communal fold.[21] The face is a site, or territory, on which various cultural and historical typographies map and remap themselves in relation to each other—sometimes erasing, other times rewriting, themselves on the one who is figured *en face*. The face of the other is thus a limit point for knowledge's knowledge about itself, becoming a mask which both conceals and congeals the enigma of alternative forms of knowledge. Accordingly, what the courts used to call "carnal knowledge" expands, via the law of libidinal economy, to include all those discourses which attempt to "know" the other through appropriation (marriage), domination (rape), observation (voyeurism), serial categorization (seduction), and so on. So, while sex is certainly one way to attempt both physical and mental access to another subjectivity, the face has the power to refuse, or rather *de*fuse, fusional models of knowledge acquisition.

This all goes to prove that "facing" the future, or our responsibilities, or the consequences of our actions, or indeed "the truth," is no guarantee of understanding that which we are facing or helping us to gauge to best response. Levinas provides a model of facing the radical Other wherein the self is but an effect of the responsibility to the Other. Deleuze, in contrast, traces ways in which we are forced to face "radicle others," a rhizome of collectivities which sometimes submit to, and sometimes resist, the culturally loaded, imperial power of the Face.

Getting Down to Business

Media-savvy corporations are currently groping their way towards facializing the consumer by harnessing computer-generated images and other digital rendering technologies. Digimask Ltd., for example, has pioneered a method whereby anyone can send in two photographs of him- or herself—one from the front and the other in profile—and then its animators get to work building a three-dimensional digital head, which can then be inserted into various virtual environments in order to give such experiences a personal touch:

> Just imagine—your head as a 3D model that talks, laughs, cries, smiles, winks, pulls faces, in fact, does everything you can do and a lot of things you can't. Now imagine your 3D head downloaded into the latest 3D game, or into a birthday e-card to send to your friends, or even to use on your mobile phone and text messages! Well, now you can with . . . a fully animatable and life-like 3D model of your head![22]

Digimask™ allows for computer-mediated communication, transcending the text-only framework which has been the staple mode of IRC (internet chat relay) and MUDs (multiuser dimensions) up to this point.[23] It also opens the door for digitally depicting standard teenage fantasies, as the company was quick to realize when launching the marketing concept of the "3D pop snog":

> Pop fans will get the opportunity of a lifetime to "cyber-snog" ["suck face" with] their favourite star at POP 2000, the largest music and lifestyle exhibition ever held in the UK. Digimasks of S Club 7 and Westlife will be projected onto massive screens at the event, with winners of an hourly competition having their Digimask beamed up in front of thousands to snog the star of their choice. . . . POP 2000, which is running for the first time this year, is tipped to be a massive hit, with a huge variety of activities to satisfy the appetite of pop-hungry punters.

Of course, not everyone can have his or her face beamed up on the giant screen, so Digimask has thoughtfully reassured potential customers, "Nobody's a loser . . . as all Digimasks™ created at the show will be emailed to their 'owners,' where . . . they will be able to use their mask to go shopping online, chat in internet chat rooms, or use on their WAP phones." The key to unlocking the corporate logic is hidden within the quotation marks that the copywriter felt compelled to use to quarantine the notion of ownership—for while once upon a time you ended up with the face you deserved, in the age of digital (not to mention surgical) manipulation, you get the face you can afford.

The possibilities don't end there. Digimask recognizes that "perhaps you want to be someone else," in which case you can build your own 3D avatar and replace it with somebody else's virtual head.[24] This option is a response to the powerful potential of "synthespians," fully rendered digital actors from films like *Final Fantasy: The Spirits Within* (Sakaguchi and Sakakibara 2001). As I write, some of Hollywood's elite actors are scanning in their physical coordinates into sophisticated animation software so that they may continue to star in blockbuster movies long after the coordinates' "owners" have succumbed to the analog inevitability of decay and destruction.

Indeed, Digimask bases its forecast profits on the premise that "human beings are vain creatures and once you get to see yourself within your PC, the online and games experience will never be the same again." The company's CEO, Gary Bracey, goes on to praise the benefits of "keeping in touch with friends and family," which "will be an entirely new experience, as you get to see clones of your loved ones from across the seas." The reference to clones is a telling one, especially when figured in connection to the lover's discourse. Certainly, the rules of the image repertoire change when the image can be so easily manipulated. Proceeding from the assumption of vanity, Digimask describes digital cloning as a much-anticipated "outlet for your alter-ego." However, the explosive narcissism of interfaciality actually has the power to dissolve rusty assumptions concerning the face as an external sign of internal individual personality.

Take, for instance, the now seemingly extinct rival company bioVirtual, which targeted its sales pitch in 2001 toward the slashing of production time, and costs, through their software package 3DMeNow Pro™. "Let bioVirtual provide your 3D human content," urges the online catalogue, explicitly assuming the interchangeability, even fluidity, of such "content":

> The application automatically generates blended morph states (morph states which are specific to the head you have modeled) for lipsynch and facial animation. The advantages of the bioVirtual system go beyond the creation and animation of the individual model: each model produced, be it a base state or a morph, *is fully mixable with any other* (as is the texture). (www.biovirtual.com, my emphasis)

Although developed according to the exigencies of production efficiency and cost cutting, such an insight into the shortcuts of designing human avatars has its ontological implications. Individuals are exposed, once again, as being completely divisible, since different elements can be mixed, matched, and cut and pasted. Here, thanks to 3DMeNow Pro™ (the sophisticated successor to Hephaestus's instruments), the human form is indistinguishable from the content. The species-being of humanity is put through the digital blender—or rather, spliced with its own simulacra—producing traces of whateverbeing in the process. (And perhaps this is a belated clue to what Heidegger meant when he wrote that "being alone is not changed by the fact that a second copy of a human being is 'next to' me" (1996, 113), a comment which plays with the notion that the other is always already a clone.)

Through the application of industry jargon such as "geometry tuning," "intelligent spline-based interface," "identical mesh typology," "visemes," and "pixel compatibility," we witness the emergence of a transitional vocabulary for the coming fusion between biological and digital engineering. Faces can now be morphed so smoothly that it is impossible to tell which features or characteristics belong to which person, as happens in Michael Jackson's attempt at racial erasure in "Black or White" (1991), a case of Benettonesque cyber-miscegenation and, more recently, in Pink's video for "Don't Let Me Get Me," in which she sings "I wanna be somebody else." Music videos thus seem to be fulfilling a Proustian prophecy, once again salvaged by Deleuze, yet soon to be harnessed by gene splicers and fetus designers:

> We can form a complex group, but we never form it without its splitting in its turn, this time as though into a thousand sealed vessels: thus Albertine's face, when we imagine we are gathering it up in itself for a kiss, leaps from one plane to another as our lips cross its cheek, "ten Albertines" in sealed vessels, until the final moment when everything disintegrates in the exaggerated proximity. (Deleuze 2000, 124)

As such, the revelations and redemptions of literature find themselves serving as the underlying code of Digimask, BioVirtual, and other corporate software. (Of course, the "corporation" is simply another kind of body.) Those of us who think within the crumbling walls of the humanities should attempt to comprehend these cultural sonic booms as they echo louder and louder off the ruins (while recognizing the fact that ruins do not necessarily symbolize an "end," as any tour operator can tell you).

Moreover, as Agamben repeatedly insists, one must look for escape routes from facialization and other daily reifications at the most saturated point: the point of the most intense and abject alienation. These technologies, which hyperterritorialize the entire body via virtual renderings of the face, also provide the starting point for creating alternatives to the pernicious equivalences between, on the one hand, the triumph of the spectacle and, on the other, the isolationist-atomic model of community (as crystallized in the "3D pop snog," for instance).

Deleuze is also quite emphatic on this point, encouraging us all to let loose "molecular populations in hopes that this will sow the seeds of, or even engender, the people to come." Indeed, "it may be that the sound of molecules of pop music are at this very moment implanting here and there a people of a new type" (Deleuze and Guattari 1999, 345–46). According to such an optimistic perspective, the teenager who utilizes avatar software in order to enact his or her adoration for a given pop star is not only unwittingly worshiping within the electronic shrine we have built to Love, Technology, and Community but forging new pathways toward the "open cosmos."

F2F, IRL

> Prepare a face to meet the faces that you meet.
>
> —T. S. ELIOT, "The Love Song of J. Alfred Prufrock"

Unsurprisingly, Agamben himself provides the shortest link between the face and whateverbeing, especially in terms of our focal triumvirate. Take, for instance, this passage from *Means without End,* where he revisits the scopic primal scene of two subjects caught in the interplay of gazes:

> I look someone in the eyes: either these eyes are cast down—and this is modesty, that is, modesty for the emptiness lurking behind the gaze—or they look back at me. And they can look at me shamelessly, thereby exhibiting their own emptiness as if there was another abyssal eye behind it that knows this emptiness and uses it as an impenetrable hiding place. Or, they can look at me with a chaste impudence and without reserve, thereby letting love and the word happen in the emptiness of our gazes. (2000, 93)

Agamben's hermeneutic system thus lies somewhere between Levinas and Deleuze, recognizing on one side that the face is produced by the appropriations of "language" (territorialization), and on the other side that the face constitutes a form of epiphany: "The face is, above all, the passion of revelation, the *passion* of language" (92).[25] For Agamben the face communicates nothing but communicability itself—the medium is the message—since "human beings neither are nor have to be any essence, any nature, or any specific destiny, their condition is the most empty and the most insubstantial of all" (94–95).[26] As such, "the face is the only location of community, the only possible city" (91).

From such a perspective, the brain-damaged subject discussed earlier was actually the only person who can see this human world "properly"—that is, beyond assumptions of propriety and identity as inscribed in the face—for this man seemed to implicitly understand a community "in which none of the visages is truer than any of the others" (99). Even Marilyn Monroe becomes an innocuous character actor. Such a unique visual-ontological perspective causes problems only in a culture where priority and emphasis are given to freezing the features into "these grotesque counterfeits of the face" (98), which contract "into an expression, stiffens into a character, and thus sinks further and further into itself. . . . [The face] turns into a grimace, which is what one calls character" (97). To approach the face in such a way is to turn the racist mantra "They all look the same to me" inside out.

The disorienting liberation afforded by wearing someone else's face—explored in films like *Seconds*, *Being John Malkovich*, and *Face/Off* and also in new-media technologies like Digimask™—points toward the situation which Agamben describes as "the possibility of taking possession of impropriety as such, of exposing in the face simply your *own proper* impropriety, of walking in the shadow of its light" (98). To be someone else "on the outside" is to appreciate better the way in which the qualities of being are in reality the directives of being-such. The desire to "be somebody," when voiced under the regime of identity, is a completely false liberation, one which actually drowns whateverbeing in the river of interpellation (while wearing the concrete-shoes of "character").

Thus, for Agamben:

> My face is my *outside:* a point of indifference with respect to all of my properties, with respect to what is properly one's own and what is common, to what is internal and what is external. In the face, I exist with all of my properties (my being brown, tall, pale, proud, emotional . . .); but this happens without any of these properties essentially identifying me or belonging to me. The face is the threshold of de-propriation and of de-identification of all manners and of all qualities—a threshold in which only the latter become purely communicable. And only where I find a face do I encounter an exteriority and does an *outside* happen to me. (99–100)

This is where Agamben's project coils into the same constellation of desire as Deleuze's, for both are eventually compelled to write under the sign of "passion," albeit a species different from the familiar image repertoire:

> Only in the black hole of subjective consciousness and passion do you discover the transformed, heated, captured particles you must relaunch for a nonsubjective, living love in which each party connects with unknown tracts in the other without entering or conquering them, in which the lines composed are broken lines. (Deleuze and Guattari 1999, 189)

Banished forever are "those glum face-to-face encounters between signifying subjectivities" (171), in favor of the erotic combinations of bodies without organs. (Consider Exhibit A for Deleuze's case: that curious phenomenon during the early stages of a courtship when the subject is undoubtedly in love and yet cannot recall the beloved's face as a totality.)

Thus, we begin to realize the extent to which other faces—not necessarily familiar, but certainly coded as "friendly"—comprise the structure of selfhood as swaddled in the world. This may go some way in accounting for the common agoraphobia which accompanies solitude and for why many of us are willing to put up with hideous people rather than be alone, without a face to shepherd our being, to help draw the boundary lines of our lives. (Otherwise, we may pitch into the vertigo of a faceless universe or drown in the pond of Narcissus.)

From this angle the obligation to constantly "be in love," or at least be searching for it, is less the cultural garnish on the biological meal—or even the social dressing on the patriarchal economic genealogy—than the universal, ethical, and existential acknowledgment of the originary "being-with" of "being-an-I." Hence, the contemporary betrayal is not the banality of infidelity but the attempted rejection of the singular-plural, and this by couples who only too willingly excommunicate themselves from society. Such legally bound or de facto couples are "singularly plural" only in a restricted sense refusing to expose themselves to the multivalent nature of the multitude—to others in general—*as if* they were a transcendental, atomistic individual. (This is the flaw with Luce Irigaray's concept of the "to be two"—specifically, its enclosed and dualistic conception, something I discuss in more detail in chapter 7.) The self-righteous, self-sufficient, and symbiotic couple is thus the new locus of Descartes's egological foundation, living in the Town-Houses of Usher, terrorized by the third-term pounding at the door at midnight.[27]

The coming community is, therefore, not a matter of throwing away the masks that hide our common humanity but of living the inessential commonality of "the I-thing encumbered with a body" (Heidegger 1996, 100). Such a community is launched from within a society "facing itself and . . . exposed to itself and only to itself" (Nancy 2000, 54).[28] *It is a refusal of Platonic re-fusing.* And perhaps most importantly, any such approach consists of keeping Kierkegaard's fan resolutely open and refusing to fold the generic singularity of beings into the singular *genera* of Being.

Ultimately, for us at least, the following questions concerning the *par-amour* become paramount. Does the *coup de coeur*—or love at first sight—somehow cut through both significance and subjectivity? Or rather, does it intensify the interpenetration of these processes into its most concentrated point? In other words, does love allow us to temporarily suspend the pernicious closures of (inter)facializing technologies? Or does it merely create an illusory access to alterity, which itself encourages the appropriation of the beloved?

FIVE

"How Was It For Me?" Not-Seeing the Non-Spaces of Pornography

> In the porn I like, everyone is satisfied, and educated, and wealthy, and contributes to worthy causes like Amnesty International, and sometimes the charity and the porn are suffused in a warm spiky glow that emulsifies their elements into a jellyish wash that engulfs and convinces even the sternest-hearted disbeliever.
>
> —ROBERT BENVIE, "The Porn I Like"

In Stanley Kubrick's final film, *Eyes Wide Shut* (1999), the protagonist, Dr. Bill Harford (played by Tom Cruise), requires a password to enter the Bletchly Manor. This password is "Fidelio." He has been banished into the night by his own jealous demons, unleashed by his wife's confession that she fantasizes about other men. The specter of the third term (in this case, a phantasmatic "man in uniform") prompts a sexual panic, and Harford wanders New York on his own libidinal odyssey, hoping to recover his wounded masculine vanity. Once inside the mansion, Harford witnesses an elaborate orgy, conducted by anonymous participants disguised by masks. In many ways this orgy is the linchpin of the narrative, as its Mogodon machinations seem to mask a far more profound secret than simply the identity of those taking part. Life itself seems to be at stake, and Harford is never sure (just as the audience is never sure) whether he is responsible for the death of the mysterious woman who pleads with him to leave.

On first release this film was considered by both the general moviegoing public and many "legitimate" critics to be disappointing. For every "hard-core" Kubrick fan, there seemed to be three people willing to condemn the film for being either boring, naïve, sexually repressed, or a combination thereof. The scene most often cited as evidence of the film's inadequacy was indeed the "much-mocked" orgy scene, described by many as "unerotic" or otherwise unappealing (as if being "erotic" was somehow its mandate, perhaps more a result of the studio's publicity machine than the logic of the narrative itself).[1]

In his study *The Shadow of Dionysus: A Contribution to the Sociology of the Orgy*, Michel Maffesoli defines the orgy as a popular means of "stating the problem of sociality or alterity" (1993, 2). He also points out that it "is necessarily disappointing." Theoretically, the orgy

> is the complete negation of the individual quality. It presupposes, it even demands equality among the participants. Not only is individuality itself submerged in the tumult of the orgy, but each participant denies the individuality of the others. All limits are completely done away with, or so it seems, but it is impossible for nothing to remain of the differences between individuals and the sexual attraction connected with those differences.[2] (129)

The most obvious difference is, of course, sexual difference, but this is something which the orgy promises to dispel in its "passional logic"—its "confusional order." On the one hand, the figure of the orgy enacts the most quotidian of experiences and, "like a subterranean switchboard, defracts into a multiplicity of effects that inform daily life" (1). And on the other, it is a liminal ritual which dissolves the invisible bonds which tether us to our social identities: "It is certain that the circulation of sexuality, the initiatory bursting of the self, orgiastic effervescence, and collective marriages all refer to the *ex-stasis*, to going beyond the individual level onto a larger ensemble" (6).

But this is where Kubrick's password, "Fidelio," becomes significant. Who are the select few allowed access to Bletchly Manor? To whom is the participant supposed to be "faithful"? Does the multiplicity of bodies merely carry the third term to a higher power, off to a potential infinity? Or does it annul the dialectical logic of the Other (into "others," the multitude)? Beyond the *meaning* of the password, the form itself is of greater significance (although it may signify nothing). The mere fact that one needs a password to join this orgy counters Maffesoli's vision of such a configuration as the figure *par excellence* of a carnivalesque, popular sociality. These are not randy peasants on ploughing night or boisterous homosocial rugby players, but the American power-broking elite. Like the ceremonies of the KKK and the Freemasons, amongst thousands of others, this orgy is a hidden articulation of political power, which creates a desiring circuit in order to channel and enhance a kind of group narcissism (hence the subservient role of the women in this scene).

Decorum and protocol are of the utmost importance, and Harford's literal "unmasking" comes from the imposture of his actions. We must therefore distinguish between different forms of the orgiastic and the various organs—and organizations—of power that they expose. Does it represent the ontological gropings of and toward an alternative form of collectivity (a "coming community," in the less vulgar sense)? Or is it a symbolic orchestration of the status quo, for the benefit of the Sadean assemblage which pushes political hierarchies into the foreground. (And the ghost of Sade should complicate any automatic equation between "popular-authentic-jouissance" and "elite-inauthentic-paranoia.")

Watching *Eyes Wide Shut*, it is difficult to isolate this scene's glossy banality from the wider orgy of the postmodern spectacle. Indeed, the masked men in tuxedos, imposing themselves on the near-naked women (all selected on the basis of a familiar aesthetic formula), leap the ever-narrowing gap dividing mainstream cinema and "adult movies." The mise-en-scène of Kubrick's orgy is reminiscent of an altogether different auteur, namely Andrew Blake, a Los Angeles pornographer who managed to make a name for himself in the wider world of 1990s erotic culture (via the [highly qualified] endorsement of "Lesbian Sexpert" Susie Bright). With his early films *House of Dreams* and *Night Trips*, Blake managed to wed that decade's particular fashion-magazine aesthetics with the hard-core formalism of X-rated pornography, creating a new subgenre based on the ancient principles of female display.

Clearly, images potentially deemed "pornographic" have always circulated beneath and between the more legitimate images produced for the open market, and these were multiplied by the invention of photography in the mid-nineteenth century.[3] Through its sheer spectral mass, the black market of pornography has continued to influence and inform the highly policed world of "art" so often that its history has been regularly punctuated by debates focusing on the practical impossibility of distinguishing erotic art from pornography (as Walter Kendrick has adequately shown).[4] For instance, were a backyard editor to take the orgy scene from Kubrick's film and splice it into one of Andrew Blake's, there would be little to disturb continuity (although some more attentive viewers may wonder why Tom Cruise is now making blue movies).

Blake's films share the ponderous, Prozac quality of *Eyes Wide Shut*, although without the suggestion of something significant lurking somewhere just out of frame.[5] Blake himself has a legion of critics, who sound remarkably like those who dismiss Kubrick's last film, basing their arguments on a lack of "action," desire, and interest. His "slight stories and wispy characterizations" irk the traditional consumer of pornographic fare, since there is very little in the way of "hot and heavy" interaction.

It is true that Blake makes "cool" movies, in the sense promoted by Jean Baudrillard (who brilliantly summed up the logic of pornography in the phrase "there is good sex somewhere, because I am its caricature" [1990, 61]). Such an effect would be impossible without Blake's large budgets, allowing the director to use 35mm film, elaborate lighting, and other "cinematic" elements eschewed by the rest of the industry, with its emphasis on output. The plots are thin, even in relation to a standard X-rated movie, and dialogue is practically nonexistent. Bizarrely, there seems to be no desire between the participants, even simulated, so that the film's libidinal economy appears to be even more alienated than the restricted economy of day-to-day commerce. This is exemplary machinic sex.

The camera pans over the surfaces of the female body with a quality which could be called "lingering" only if the camera operator were an android, with little understanding of the more affective effects of voyeurism. There is no sweat or saliva, and very little of that most vital of fluids, sperm (especially after 1993, when Blake practically stopped using men altogether). Hence, these films fetishize the Body-without-Fluids of the high-AIDS era. They portray desiring circuits without desire and dovetail with the particularly hyperreal articulations of the supermodel phenomenon. If Blake did not directly influence the orgy scene in *Eyes Wide Shut*, then he certainly did indirectly, through the deployment of a particularly 1990s aesthetic which wedded the soft-bondage output of Helmut Newton with the emotionless poise of the Milan catwalk.

In his interview with Susie Bright, Blake states, "I still can't believe how much I've been copied, but I don't think I've ever been duplicated. They always seem to miss the mark on the production design, the choice of location or the styling. It comes down to taste. They don't have the vision or the concept of beauty." Indeed, Blake sees his films as part of a crusade against the sausage factory approach to pornography which has dominated the industry since the late 1970s, referring to the rival studios as "sexual Muzak." While admiring his attempts to widen the horizons of the genre, Susie Bright takes issue with Blake's ornamental evacuation of desire, stating that she would rather see a woman "begging for it in a ditch" than enduring the intense aestheticization which has always set limits on the definition and depiction of "female sexuality." In response, Blake simply pleads personal taste and correctly points out that women enjoy his films as much as men (which of course does nothing to counter Bright's point).

A typical Blake scenario unfolds as follows: A beautiful woman discovers either a fetishistic object (such as an ice phallus or Japanese duo-balls) and/or an arousing liminal environment (almost always an empty modernist mansion). She then proceeds to succumb to the pleasures of onanism, itself always a gesture deeply imbricated in the prosthetic (that is, technological), even when only involving the hand (according to Stiegler's work on the hand and grasping, Lacan's understanding of the phantasm as instrumental, and Rachel P. Maine's work on the vibrator as the "technology of orgasm"). In the tradition of Hitchcock's *mise-en-abyme*, the viewer spies on someone (usually another woman) spying on this first woman. This technique deflects the gaze into a house of mirrors, both disavowing and multiplying the "guilt" of watching, as well as embodying the essential third term. The editing seems willfully monotonous, building to a climax while simultaneously short-circuiting any such possibility through its asymptotic approach.[6] More and more bodies may enter the frame, mimicking the overdetermined tableaus of all pornography, yet, as we have already seen, this orgy is "necessarily disappointing."

The fact that a pornographer can couch his work in semilegitimacy, via the notion of the auteur, owes much to the cultural exchange between forms and genres—for instance, the bizarre cross-pollinations between the French New Wave and Hollywood titillation movies like *9½ Weeks* and *Wild Orchid*.[7] More recently, editorials around the world are debating the "two-way traffic of codes" between fashion and pornography through the work of photographers such as Terry Richardson, Mario Testino, and Juergen Teller. Former porn directors are now trying their hand at more mainstream material, such as Gregory Dark, who moved from *New Wave Hookers* to Britney Spears videos (not such a quantum leap, really).

This traffic also goes in the other direction, most spectacularly, perhaps, in the case of the former deputy chief of the Australian Censorship Board, David Haines, who decided to stop banning such movies in favor of making them himself. The point not to be missed, however, is that this crosstown traffic flows according to the *same* system. Pornography may be the ghetto of the art world, but it rides the airwaves of the same economic vectors. The depictions of sex in both Kubrick and Blake are saturated by capital and bespeak the erotics of reification. (Indeed, the latter's historical influence may lie less in the ripples which reached Kubrick's film set than in the way he managed to generate so much money while simultaneously banishing the "money shot" from his productions.)

But why resuscitate the exhausted category of pornography, especially in relation to the equally exhausted category of art? Simply because Blake's soft automatons give us yet another glimpse of the contours of whatever-being. If we suspend the abundant problems associated with the industry itself (most of which are shared by the up-market sets of Hollywood), these particular films delineate the non-space of desire, tracing the evaporation of a familiar mode of *jouissance* in the era of highly coded and commodified spectacle. Kubrick's film takes this phenomenon and attempts to reinject the secret, or the accursed share of enigma. In *Eyes Wide Shut*, Thanatos reenters the picture, forcing Eros to take stock. In Blake, however, it is the very exhaustion of the discourse of sex which gives rise to the "truth" of its form and function (just as an exhausted person senses the troubling "truth" of being without distraction).[8]

Baudrillard had his finger on this particular pulse when he stated that pornography's "ghostly ambiance comes to it from its anticipation of dead sex in living sexuality, from the weight of all the dead sex (as one used to speak of the weight of dead labor on living labor). In so doing, porn makes sexuality appear superfluous—that is what is obscene: not that there is too much sex, but that sex is too much" (1990, 32). Writing against the disingenuous romantic discourse which seeks to distinguish love from lust, Baudrillard claims:

> To love someone is to isolate him from the world, wipe out every trace of him, dispossess him of his shadow, drag him into a murderous future. It is to circle around the other like a dead star and absorb him into a black light. Everything is

> gambled on an exorbitant demand for the exclusivity of a human being, whoever it may be. This is doubtless what makes it a passion: its object is interiorized as an ideal end, and we know that the only ideal object is a dead one. (105)

Through its relentless mode of accumulation, where the actors move between partners as if at a smorgasbord, standard pornography negates that annexing of the Other which almost always occurs in the name of love. In Deleuzian terms, pornography deterritorializes love. The bulk of this genre's frantic accumulation thus spasms into its opposite without any perceptible reversal, into a potlatch of expenditure. In contrast, Blake's movies hover between the two tendencies: at one moment circling around the female body "like a dead star," and the next slipping metaleptically from one body to the next; fetishizing none by fetishizing all.

Hence Baudrillard's observation that "There is, then, a kind of love that is only the froth of a culture of sex, and we shouldn't have too many illusions about this new apparatus of ambience" (102).

Working Ass in the Age of Biomechanical Reproduction

> True pornography is given us by vastly patient professionals.
>
> —THOMAS PYNCHON, *The Crying of Lot 49*

We have already seen how Robert Wall offers the "character actor" as a possible figure through which we can think whateverbeing. Agamben himself believes that porn stars (provided they are not really "stars") also embody this particular mode of anticipation. In the chapter in *The Coming Community* entitled "Dim Stockings," Agamben states that pornography, along with advertising, "escort the commodity to the grave like hired mourners" and are thus "the unknowing midwives of this new body of humanity" (1993a, 50).

He continues:

> The commodification of the human body, while subjecting it to the iron laws of massification and exchange value, seemed at the same time to redeem the body from the stigma of ineffability that had marked it for millennia. Breaking away from the double chains of biological destiny and individual biography, it took leave of both the inarticulate cry of the tragic body and the dumb silence of the comic body, and thus appeared for the first time perfectly communicable, en-

tirely illuminated. The epochal process of the emancipation of the human body from its theological foundations was thus accomplished in the dance of the "girls," in the advertising images, and in the gait of fashion models. This process had already been imposed at the industrial level when, at the beginning of the nineteenth century, the invention of lithography and photography encouraged the inexpensive distribution of pornographic images: Neither generic nor individual, neither an image of the divinity nor an animal form, the body now became something truly *whatever*.[9] (47–48)

Agamben—taking his cue from Siegfried Kracauer—thus recognizes a techtonic, or epochal, shift in the "perfectly fungible beauty of the technologized body," noting that "a promise of happiness" accompanies the spectacularized body: "To grasp a whateverness one needs a photographic lens" (47, 49).

Just as the techniques of scientific and medical research and administration overlap the different spaces of modernity (the gynecologist's office, the concentration camp, the pornography set), the various blueprints of whateverbeing seek to escape the intense organization and codification of "bare life":

The mortality of the organic body has been put in question by its traffic with the body without organs of commodities; the intimacy of erotic life has been refuted by pornography. And yet the process of technologization, instead of materially investing the body, was aimed at the construction of a separate sphere that had practically no point of contact with it: *What was technologized was not the body, but its image*. Thus the glorious body of advertising has become the mask behind which the fragile, slight human body continues its precarious existence, and the geometrical splendor of the "girls" covers over the long lines of naked, anonymous bodies led to their death in the *Lagers* (camps), or the thousands of corpses mangled in the daily slaughter of the highways. (50, my emphasis)

Whateverbeing is thus *post*-structuralist insofar as it recognizes the radical interchangeability of individuals in a system far larger than the subjects which inhabit it *but does not reduce the individual to this predicate*. Here the "girl next door" could in effect be *any* girl next door but *in fact* is this one (in her *suchness*, to wax Heideggerean). Hence the fundamental tension of techtonics: How to reconcile the anonymity and interchangeability of the Other, or indeed the Self, with the recognition of *absolute irreplaceability* (a repressed subtext which lies at the heart, so to speak, of the lover's discourse itself)? In an Andrew Blake film, concepts like "random" and "fated" seem even more irrelevant than in traditional pornography—which all conspire to suggest that the secret of "hard-core" pornography is that there in fact *is* no core.

Pornography thus points to this "new body of humanity" through the complete exhaustion of the system's own logic. In a strategy favored by Slavoj Zizek, the subject takes the ruling ideology at face value, believing in its dictates to an unreasonable degree, beyond even those who are the architects of its ethical pressures, thereby sabotaging the system through the bluffer's game of chicken. No ideological regime functions on "blind" subservience alone, but rather operates through the flexibility of a certain "ironic distance," which itself is easily reappropriated into the wider state-media-industry apparatus. The only way to expose the various violences of the system is therefore to collapse this ironic distance and stubbornly refuse to play the winking game. In this sense, the spectacular reification of the body is "the danger that saves," and has the potential to clear the path for its liberation from centuries of surveillance and control, whether this be in the name of God, public order, hygiene, or moral character.

Pornography is thus the maximum intensification of capital *qua* the human and is therefore a saturation point or "black hole." As with quantum physics, it is impossible to see the movement of this immanent reversibility, and yet it definitely occurs. We know this because advertising sells us an identity (or a set of identities). Pornography, on the other hand, *sells our identity*. By completely severing the sign from the body, pornography has the potential to break with the interpellative power of the market: as midwife, as hired mourner. (Baudrillard again: "Right at the very heart of news, history threatens to disappear. At the heart of hi-fi, music threatens to disappear. At the heart of experimentation, the object of science threatens to disappear. At the heart of pornography, sexuality threatens to disappear" [1994, 6].)

Pornography is advertising taken to the nth degree. Where the latter launches itself toward a demographically delineated consumer, the former hurls itself into the less codified landscape of the libidinal economy itself. Where advertising depicts caricatured perky housewives and disgruntled teenage children, pornography depicts the caricature of such caricatures, straddling the line between the stylized and the abstract. In other words, porn stars do not "represent" a type but function in relation to an auto-affection machine which—ultimately—produces an inkling of whatever-being. As a mobile set of generic effects, pornography follows its own technological trajectory. Narrative—and even characterization (no matter how tokenistic)—evaporates, leaving an essence of an originary relation. And it is this essence, as ridiculous as it may sound, which beckons us toward a redefinition of community, one which emerges *on the other side* of reification, exploitation, sexual violence, and profit.[10]

Certain Marxist or feminist voices may argue that this essence is merely an industrial by-product, like the chemicals siphoned off into rivers by pharmaceutical companies. But this would miss Agamben's point that "While commodification unanchors the body from its theological model, it still preserves the resemblance: *Whatever is a resemblance without archetype—in other words, an Idea*" (1993a, 48). What we have been calling "essence" is irreducible to a political program, or to the basis of something we may be tempted to call "the human," with its attendant historical and metaphysical baggage.

Where advertising uses sex to sell products, pornography harnesses sex to sell something the body already has, or is at least theoretically capable of doing (whether we treat "sex" as a noun or verb). This perverse arrangement relies on the eroticization of mediation itself, since the consumer would not "enjoy" pornography without some kind of twist on that which is, after all, "ready to hand." (The argument that only those who can't get laid rely on pornography simply does not stand up.)[11] Indeed, Blake's movies are nothing if not homages to the "aura of frozen intangibility" (Agamben 1993b, 42) of commodified technology itself. Even *Eyes Wide Shut,* to all appearances a rather old-fashioned film, unconcerned with "technology" as such, obsesses on the phantasm as a psychic technology underpinning all social relations in the contemporary world (what I have elsewhere called "the goat in the machine").

If "making love" is the linguistic hinge between the lover's discourse and the sexual act, then Zizek is correct in pointing out the necessity of at least "a minimum of phantasmic narrative as a symbolic support" between lovers. This is based on the psychoanalytic assumption that "any contact with a 'real,' flesh-and-blood other, any sexual pleasure that we find in touching *another* human being, is not something evident but inherently traumatic, and can be sustained only insofar as this other enters the subject's fantasy-frame" (1997, 65, 183–84). Dr. Harford no doubt has a strong investment in believing in an "authentic" congress between his own body and his wife's, and he is shattered to find out a basic Freudian truism concerning the way people navigate the gulf between Self and Other—especially when this gulf has been physically breached. Love thus threatens to spasm into jealousy at any moment, just as the Maffesoli orgy constantly runs the risk of being reterritorialized into "the gangbang."

And it is the artificial eye of the camera which best captures this struggle.

Beneath the Mosaic: I.K.U. and Orgasm Coding

Are the techtonics of postmodern porn part of a global complex? Well, yes, insofar as aesthetic mutations are rarely, if ever, limited to geopolitical borders. Certainly, to speak of "pornography" as a monolithic genre, as I have up until now, is to neglect all the internal inconsistencies and nuances which lurk beneath every label. But just as we can talk about something called "music" while simultaneously acknowledging the incompatibilities of, say, jazz and pop, we can also consider pornography a category traversed by many contradictory interests. Certain formal relations are shared on the level of structure, and this assemblage called "pornography" (conceived as the set of all possible "pornographies") manages to cohere around an originary techno-erotic relation.

Having said this, there is little doubt that Andrew Blake's films are relentlessly WASPish, especially in contrast to the videos in hip-hop's concerted attempt to turn MTV into a hard-core porn channel. If *House of Dreams* or *Night Trips* is "American," it is through a dominant genealogy which extends back through Hugh Hefner's playboy bunnies to the Ziegfeld Follies, rather than to the less visible Coochie Dance or Harlem burlesque. Blake's actors are uniformly white, and his props are constantly coded in terms of the luxury class. Nevertheless, all of pornography's various *incarnations* share a certain structure concerning the voyeurism, mediation, fetishism, and metageneric identification discussed above.

Blake's films take place in the "non-places" of Marc Augé's supermodernity (1995), within a Los Angeles both everywhere and nowhere, overpopulated and deserted. An excess of natural light is key to this director's "vision," and the architecture of his environments exemplifies a perfect mesh of Charles Eames and Jeremy Bentham, resulting in a kinky modernist panopticon. Such locations evoke Deleuze's "any-space-whatever," which constitutes "a perfectly singular space, which has merely lost its homogeneity, that is, the principle of its metric relations or the connection of its own parts, so that the linkages can be made in an infinite number of ways. It is a space of virtual conjunction, grasped as pure locus of the possible" (Deleuze 1986, 109). In Blake's world this serves as a fancy way to signify "wealth" in its most stylized and vacant aspect, connecting to a global grid of airports, five-star hotels, and other indistinct environments for *Homo generica*. Indeed, Blake's Los Angeles represents the other side of the same postmillennial coin stamped with Ridley Scott's *Blade Runner* (1982): the neo-Byzantium-cum-Gothic future-noir aesthetic, which has had a stranglehold on our expectations for twenty years now. Between these two styles lies the ever-shifting terrain of art-masquerading-as-porn-masquerading-as-art.

Taiwanese-born American artist and "digital nomad" Shu Lea Cheang has thrown down the gauntlet to Blake and his imitators with her film *I.K.U.* (2000). Described by the director as a "Japanese science fiction pornographic" concept movie, *I.K.U.* seeks to defy the repetitive formula of mainstream triple-X movies, and tells the tale of the giant Genom Corporation, which is established in 2016 and specializes in "digital desire entertainment." What little plot there is revolves around the company's advanced new I.K.U. chip, which "aims at sending sexual pleasure signals collected from I.T. and Genom technology, directly to the brain without physical friction."[12] Such pleasure signals are first gathered by *Blade Runner*–esque sexual androids, deployed throughout the New Tokyo community in the interests of "orgasm hunting," for a gigantic database of sexual preferences.

According to the Japlish information on the film's website, "Data [is] collected and put to life by a replicant called Reiko. . . . The wave signals obtained through the eye and ear, is calculated to reach the brain. Access from the user, transmits IKU data through the server on demand, in a coded pattern form. The data is then decoded by an IKU chip, enabling the user to audio visualize the data."

Cheang insists: "I did not make this film for masturbation. I made it for collective orgasm. It is like doing drag, like transsexuality, the way we are interfacing with technology and extending our identity into it" (2002). For its part, the now-defunct style bible *The Face* claimed that "it is this intersection of sex and technology that lends *I.K.U.* its groundbreaking pedigree—and not just as pornography. *I.K.U.* is one long lavish visual metaphor for the sexual freedoms afforded by the internet, fantasies you can indulge with others regardless of gender, social constraints or even physical possibilities." If we graciously ignore the historical and ontological ignorance supporting this review, and indeed much of this film, we can skip straight to the issue of separating physical and virtual sex—specifically, Cheang's question "What will the human race create from the new pleasures gained after being free from the ecstasy gained by physical friction?" (The question should perhaps be reversed: "What ecstatic pleasures will create a new human race?")

Cheang's most interesting point, however (not really explored in the film itself), relates to the eroticization of technologies developed especially to censor the depiction of sex:

> The I.K.U. data is made up of visual and voice data in mosaic. Of the countries which have regulations against sexual expression, Japan's is one of the most nonsensical. Hiding genitals by mosaic peculiarly grew in the field of pornography in Japan. Japanese men have gained their ecstasies by imaginating women's genitals beyond the mosaic visuals and voices. That is to say that Japanese people have made themselves into an exceptional nation which is aroused by a mosaic. (2001)

There is little doubt that this form of fetishization is yet another techtonic development in terms of the multidirectional tug-of-war over the libido.[13] Indeed, there would be much to say on the topic of digitized obfuscation and revelation if Baudrillard had not already said it. But the important thing to note in terms of the previous discussion is the extent to which the libido is itself a technology, a prosthesis in which we navigate the world and "externalize" ourselves among other externalized beings. And as we have seen, what we call "love" is (merely?) the discursive channeling of this libidinal *techne.*

Through half-hearted warnings concerning the hazards of technologically mediated sex, *I.K.U.* could be accused of falling into the familiar trap of romanticizing "authentic" intimacy. The implicit claim is that mediated sex, or "repetitive pornography," is dehumanizing or, at the very least, depersonalized. As I argued above, though, vertiginously dull pornography may in fact open the hazardous route to an alternative way of thinking humans-as-people. (This doesn't mean we need actually watch it, since it suffices that the symptom exists.)

But while this route is usually blocked by the dictates of narrative cinema, we can find signs pointing to an exit.

Take, for instance, Roman Polanski's film *Bitter Moon*, in which the male protagonist ruminates over a particularly debauched period of his life. Speaking of a one-night stand (merely one of many), he confesses, "In her eyes, I could see the eyes of the next woman." Such an attitude—or, to be more accurate, perception—speaks of the metaleptic movement of libidinal flows. ("Metalepsis" is defined by the *Oxford English Dictionary* as "consisting in the metonymical substitution of one word for another which is itself figurative"; we ourselves need only replace "body" for "word" in this context.) Here we encounter Lacan's "desire that is based on no being—a desire without any other substance than that assured by knots themselves"(1999, 126). Certainly, in Polanski's example at least, this perception is compromised by the chauvinist lenses which filter it. We should not be too quick, however, to rule out the possibility that the lover's discourse (which demands individual, proper essences) often disguises itself as moral outrage ("bad, sleazy man") *precisely in order to deny the existence of an ex-static essence.* Such a Deleuzian essence is, as we have seen, beyond or behind those who incarnate it, and yet it is never completely severed from individuation itself.[14]

Such a slippery form of lust has nothing to do with attempting to capture the essence of Woman as a transcendental category—the familiar view bequeathed by Romanticism, consisting of an *ewig Weibliche* in which each individual woman participates. Indeed, during his lectures on feminine sexuality, Lacan made an effort to be clear on this point: "There's no such thing as Woman, Woman with a capital W indicating the universal" (72).

Moreover, in an earlier lecture published in the same text, Lacan makes a decisive statement leading to this conclusion:

> Don't you see that what is essential in the feminine myth of Don Juan is that he has them one by one? . . . From the moment there are names, one can make a list of women and count them. If there are *mille e tre* of them, it's clear that one can take them one by one—that is what is essential. That is entirely different from the One of universal fusion. (10)

The computer industry has invented a term perfectly adapted to this phenomenon: "hot-swapping," whereby one device can be removed from a computer port, and another can be attached, *and recognized*, without rebooting. We would be hard-pressed to find a more succinct example of the logic of pornography and, by extension, the prosthetic relation to Otherness in the plural, serial sense. Sex is hot-swappable when it adapts itself not only to the acknowledgment of the third term but to its material presence. The orgy may begin with four, but society begins with three (or, more accurately, the movement from two to three).

Indeed, the world of information technology provides us with yet another metaphor for this *original* metaphor that we have been pursuing to the point—and hopefully beyond the point—of exhaustion. Called "circle-jerking," it is the technique currently favored by pornographic and other commercial websites in which the attempt to close one browser window only succeeds in opening a multiplicity of extra windows, which themselves advertise other websites. Should you try to close these new pop-up menus, the situation only becomes more vertiginous, as browser after browser opens up on the desktop, revealing a *mise-en-abyme* of windows within windows within windows. The machine itself seems to be suffering from a state of exponential panic, the kind which can arise in the non-space of a finite self in proximity to a potentially infinite set of others. Sex itself partakes of this panicky proliferation, as the physical and metaphysical borders are temporarily effaced and redrawn, allowing a glimpse of the technological artifice which sustains the autonomous, sovereign subject. (Hence Don Juan's need for linear, serial—that is, *non-orgiastic*—conquest.)

As we saw with Kierkegaard's diary, true seduction plays within the mirrors of this knowledge, in contrast to the essentially narcissistic, self-bolstering antics of the crude Casanova approach. It is therefore a profound irony that pornography is aligned with the former orientation. For while the consumers of pornography may be largely composed of would-be Don Juans, fragile egos simulating serial sexual conquests, the geometry of the "narrative" works against such an ingrained phallic habit. The naked body here function as a synecdoche with no Ur-referent, just as Polanski's scenario depicts adultery with no focus—the "becoming-everybody" which undermines the interiority of fetishism:

> As the saying goes, desire is an infinite metonymy, it slides from one object to another. In so far as desire's 'natural' state is thus that of melancholy—the awareness that no positive object is 'it,' its proper object, that no positive object can

> ever fill out its constitutive lack—the ultimate enigma of desire is: how can it be 'set in motion' after all? How can the subject—whose ontological status is that of a void, of a pure gap sustained by the endless sliding from one signifier to another—none the less get hooked on a particular object which thereby starts to function as the object-cause of his desire? How can infinite desire focus on a finite object? (Zizek 1997, 81)

In the hot-swappable, circle-jerk of metalepsis, there is nothing for the libido to "catch" on: It is pure movement. And wherever there is pure movement, we are already in the realm of the posthuman.

Inter-Passivity

Zizek himself mines the world of information technologies for keywords which reveal more than they intend, even coining his own term, *inter-passivity*. Presumably the opposite of *interactivity*, this term denotes those abundant mechanisms we employ to avoid actually doing something. Hence the hidden function of canned laughter—namely, that it laughs *for* us, so we don't have to feel obliged to respond to the program in an active mode, expending unnecessary energy laughing for ourselves. Such examples could be multiplied: video machines that watch television for us, photocopiers that read books for us, electric muscle stimulants that exercise for us, animated talking paper clips that write letters for us, holiday programs that travel for us, lifestyle programs that live for us, and even post-feminists who objectify their bodies for "us" (i.e., the male gaze). All this in order to leave us time to . . . do what, exactly? Daily existence thus begins to resemble that old tautological toy in which you press a button, causing a mechanical hand to appear from inside the box and turn itself off again. (Yet another perfect metaphor for the founding metaphor of sex.)

On one level, pornography is part of this constellation, relieving us of having to have acrobatic, spectacular sex. But on another, it enacts the "machinic enslavement" of a libidinal economy almost completely penetrated by the commodity economy. In this latter sense, other people, our "sexual partners," are merely "workstations" on which we perform. The women work on the men like industrial vacuum cleaners, while the men "service" the women in the mode of mechanical pistons, resulting in "a kind of vaguely co-ordinated agglomerate" (Zizek 1997, 180).

"This change of the body into a desubjectivized multitude of partial objects," argues Zizek, results in an assemblage comprised of "organs without a body," "machines of *jouissance*" (ibid.). The task, however, is not to merely dismiss pornography as the apex of the instrumentalizing of the human body qua capital, but to trace the new network of valences which arise from such an intense reification:

> It would be interesting to approach, from this paradox of interpassivity, Schelling's notion of the highest freedom as the state in which activity and passivity, being-active and being-acted-upon, harmoniously overlap: man reaches his acme when he turns his very subjectivity into the Predicate of an ever higher Power (in the mathematical sense of the term), that is, when he, as it were, yields to the Other, 'depersonalizes' his most intense activity and performs it as if some other, higher Power were acting through him, using him as its medium—like the mystical experience of Love or like an artist who, in the highest frenzy of creativity, experiences himself as a medium through which some more substantial, impersonal Power expresses itself. (126)

It thus becomes necessary to map the emergent syntax from within the white noise of e-jaculation, or what we have hitherto been calling essence. In terms of the lover's discourse, this is less "I'm coming" than "*Something* is coming through me."

I.K.U. is pronounced "ai-kei-ju," which is the Japanese equivalent of the English phrase "I'm coming." As already noted, it is surely by no means an accident that Agamben offers pornography as a herald of his coming community, and one need not labor the logic which dwells within the innuendo. The important thing to remember, however, is the particular temporality of the *arrivant(e)*, which comes either before the orgy or after the orgy, but never during (*intra-festum*). On the linguistic level, in distinction from the biological, we never actually arrive. While having sex, we are either coming ("I'm cooommminnnggggg"), or we have just come ("So, how was it for you, baby?"). Orgasm is either future tense (even during the overwhelming presence of the spasm itself), or it is past tense . . . gone, extinguished, history. Hence, "I" am merely an effect of that which never arrives and yet has already left.

It is precisely in these terms that we must remind ourselves that Agamben's community is not a utopian program to be realized, since this would recuperate it within the history of prophecy from which it seeks to escape.

Deleuze and Guattari have also unpacked the differences between subjection and subjectification within this technical milieu. They make a distinction between machinic enslavement and social subjection according to the following criteria:

> There is enslavement when human beings themselves are constituent pieces of a machine that they compose among themselves and with other things (animals, tools), under the control and direction of a higher unity. But there is subjection when the higher unity constitutes the human being as a subject linked to a now exterior object, which can be an animal, a tool, or even a machine. The human being is no longer a component of the machine but a worker, a user. He or she is subjected *to* the machine and no longer enslaved *by* the machine. (1999, 456–57)

They go on to say that in machinic enslavement, "there is nothing but transformations and exchanges of information, some of which are mechanical, others human" (458). But this very process undermines an increasingly "artificial" distinction between the human and the machine. This is not part of a grand narrative where an a priori humanity is increasingly absorbed by the technology it produces, but rather a process in which subjection and enslavement "constantly reinforce and nourish each other," reproducing the abstract machines from which humans are an inherent effect.

Once again, it is not a matter of juxtaposing flesh with hardware in order to contrast human with machine, since technology is immanent to the human, and vice versa. (In *Lady Chatterley's Lover*, Oliver Mellors bluntly states, "What is cunt but machine-fucking!" (Lawrence 1994, 217), without realizing that such loaded words could be turned against the neo-Luddite-humanistic location from which they were spoken.) The question to be asked at the beginning of a new century is both depressingly similar, yet fine-tuned in directions which would have both D. H. Lawrence and his critics spinning in their graves, this being: Has sexual pleasure been appropriated by the Spectacle to the extent where every moment of life-affirming *jouissance* manages to be preempted, framed, and/or re-funneled back into the machinic functions of the restricted economy? Can the BwO (not only the Body-without-Organs but also without-Organon and -Orgone) enter any space untouched by this drowning flood—that is to say, by the increasingly liquid logic of reification?

Woody's Wood Pecker; or, The Life of an Ordinary, Sensuous Thing

> In Scotland, two people were discovered to have begged surgeons to amputate healthy limbs. Kevin Wright said of his left leg: "I just didn't want it. It didn't feel part of me." The operation was performed at the Falkirk Royal Infirmary.
>
> —*The Observer*

The porn industry relies on a steady supply of what is referred to in the trade as "woodsmen." These are male actors who can provide "good wood," who, in other words, have the ability to maintain erections throughout demanding shooting schedules and the pressure to perform on cue. This term refers less to the virile image of a lumberjack hacking his way through a forest of young women than to the organ itself: the penis as phallus, as a wooden-tool symbol.

In one of the most oft-quoted passages from Karl Marx's *Capital,* we read:

> The form of wood . . . is altered if a table is made out of it. Nevertheless the table continues to be wood, an ordinary, sensuous thing. But as soon as it emerges as a commodity, it changes into a thing which transcends sensuousness. It not only stands with its feet on the ground, but, in relation to all other commodities, it stands on its head, and evolves out of its wooden brain grotesque ideas, far more wonderful than if it were to begin dancing of its own free will. (1977, 163–64)

In fact, the erection of the woodsmen does, in a sense, "dance of its own free will," provided we recognize the ironic or qualified sense in which Marx employs this image. The spectacular penis (or, indeed, vagina) "transcends sensuousness" to the degree that it is now part of an exchange economy. However, this mode of exchange (and the value it produces) predates capitalism by many thousands of years, beginning with the emergence of society itself and the forms of kinship and alliance which support it. Pornography forms a limit for the organ-ization of alienation that Marx speaks, as far as it strips the body of its transcendental residues (the point emphasized by Agamben cited earlier).

Here we return to Baudrillard's observation that the pornographic aura "comes to it from its anticipation of dead sex in living sexuality, from the weight of all the dead sex" (1990, 32). Phallocentrism reaches its peak in the mad rush for Viagra, a piece of biotechnology so advanced, and so in tune with the times, that it can isolate the physiological triggers which control blood flow to the penis and bring them to a point of acute "standing reserve." Such an exquisite scientific accomplishment not only anticipates the coming revolution in nanotechnology (which will do more to erase the distinction between "outside" and "inside" than every tome on deconstruction put together) but completes the historical trajectory of prosthetic grasping. In pornography the penis does not grow a brain, à la Marx's table, but rather a fist, in order to aggressively assert the authority of its owner. (Need I refer once more to the famous scene in Kubrick's *2001* [1968] in which a bone-tool-weapon becomes a spaceship-tool-probe?)[15] Such a social configuration leads to the situation described in the news item introducing this section, in which someone can feel completely ambivalent, even hostile, toward a crucial part of his own body. It is similarly inscribed in the "meat puppet" mentality of much cyberpunk and the corporate rhetoric which followed in its wake. Even the ubiquitous funk lyric "Shake your money-maker" reduces the body to those parts which can be used to turn a profit.

But is it really a straightforward case of reduction, pure and simple? Surely, we would be ignoring the previous discussions if we talked in such hackneyed, unilateral terms. Where ten years ago girlie calendars on the garage wall depicted airbrushed young women draped over cars, they now show exotic cyborg-hybrid creatures: half-woman, half-car, waxed, polished, and digitally enhanced. Clearly, in terms of traditional feminist politics, this is hardly progress. However, just as definitions of humanity were reforged in the crucible of the industrial revolution and on the anvil of World War I, they are now being reconfigured and recoded by the wider digital economy. The *to-be-looked-at-ness* of women continues to bear (and bare) the brunt of these redefinitions; nevertheless, Agamben sees more than merely a dim hope in the stockings of these "midwives of a new humanity." For whether we place our (admittedly shaky) faith in the laws of reversibility, or deterritorialization, or the accursed share, such hyperobjectification will—inevitably—provoke questions about the status of such objects in relation to the subjects who supposedly master and market them. This is the revenge of the object—all the sweeter for being served cold.

Speaking historically, we can point to epochs where machinic enslavement replaces social subjection, and vice versa. Our era is unique, however, in the particular constellation under discussion: the ubiquity of the spectacle (advertising, branding, pornography), the sheer mass and mobility of the multitudes (overpopulation, immigration), the exhaustion of the political (the "withering" of the state, the "death" of grand narratives and counter-narratives, etc.), and the apotheosis of alienation (in the "media" and its various moral forums, such as romantic comedies, soap operas, news reports, and talk shows).

It is in the aptly named *Eyes Wide Shut* that Kubrick made his final comment on our intensely scopophilic epoch, shedding new (electric blue) light on the relationship between fantasy and archetype. More precisely, *Eyes Wide Shut* patiently unpacks the ways in which phantasmatic archetypes function as a *technique* which sometimes undermines, other times reinforces, the discourse of exclusive love. What's more, it does so using a visual grammar borrowed from the Los Angeles pornographers who continue to produce such phantasmatic archetypes for private consumption. And this is why Kubrick's Alice (Nicole Kidman)—much like Carroll's—is first introduced to us in front of a mirror, reflecting the uneasy alliance between self-fashioning (Foucault's "technologies of the self") and the intrinsic spectacularity of the image repertoire.

For while it is the theorists such as Foucault and Luhmann who illuminate the forces driving the cultural coding of intimacy—in this case, the sociohistorical aftershocks of forcing the libido (desire), the law (marriage), and love (fidelity) to share the same bed—it is up to the artworks of each era to absorb and adapt to these self-same techtonic movements. Despite the different attitudes that accompany each epoch, however, the spectral code of "love" continues to function as the key term, itself an operating system which no longer seems to be able to locate the file "community" on which it was originally designed to operate.

SIX

A Self of One's Own?

> Only where man is essentially already subject does there exist the possibility of his slipping into the aberration of subjectivism in the sense of individualism. But also, only where man remains subject does the positive struggle against individualism and for the community as the sphere of those goals that govern all achievement and usefulness have any meaning.
>
> —MARTIN HEIDEGGER, "The Age of the World Picture"

> He asked her one night
> which one of herselves was sleeping with him.
> And all she could reply was: this one."
>
> —CARRIE OLIVIA ADAMS, "Vermilion" in *A Useless Window*

> Our identity is always a case of mistaken identity.
>
> —MELANIE KLEIN, in A. S. Weiss, *The Aesthetics of Excess*

Consider the little games lovers play with each other. You know the kind: "Would you still love me if I was in a horrible accident and lost both my legs? Or if my face became paralyzed and I always spoke like *this*? Or if I lost my job and had to clean toilets? Or if I suddenly started wearing those jeans with the transparent pockets we saw yesterday in that shop window?" The perfectly reasonable assumption is that beneath this litany of horrors is a unique being worthy of being loved, a being immune from physical distortion or fashion disasters. But what is it exactly that we love about this being, if we cannot confidently locate a definable essence? Are we happy to just answer, "Their soul"? Some kind of intangible kernel of being? If so, then how do we assimilate the trauma of the second love? How do we account for the insult of multiplicity, as discussed in the previous chapter?

The ever reductionist Alain de Botton sums up the problem quite succinctly in his book *Essays in Love*: "Once we locate beauty in the eye of the beholder, what will happen when the observer looks elsewhere?" (1994,

101). Clearly, love is not simply the harmonious blending of two complementary contents—the contents of our unique individualities. We could not find a less appropriate symbol than the combination of yin and yang to represent the subtextual decodings of the lover's discourse, since the complementarity is too Neoplatonic, too complicitious with the discourse itself.

"There but for the grace of God go I," states the folk wisdom, recognizing the invisible contingencies which separate us from our neighbors (an aphorism which shares a rhetorical kinship with the deceptively profound "If it's not one thing, it's another." A lottery ticket, a different science teacher, an allergy to spaniels, an unassuming foreskin, and, indeed, the Grace of God itself may be the only things which we can point to as the bases of distinguishing Tom from Dick from Harriet.[1]

The crucial move, however, is to make this statement stand *without* relying on the naïve onto-liberalism that would posit a shared substance, specifically "humanity," which is only then, through life's trials and tribulations, imprinted with "difference" (i.e., identity). The trick is not to advocate a *Homo generica*, or Home-Brand Humanity, which exists a priori to experiential grids such as ethnicity, class, and gender, only then to be shaped through time. Instead, we must give serious thought to a constitutive difference at the very origin (Stiegler would say "default") of being, which *itself* is the medium of transmission—to share the fact that we have nothing to share, not even our somatic structures, since these can be radically revised through birth, injury, surgery, or design.[2]

This is all to say that some very complicated maneuvers are being made whenever we utter the personal pronoun "*I*." Indeed, the mysterious inconsistencies of deixis challenge the authority of Descartes's "I think, therefore I am" by stalling any self-evident logic at the dictum's very first word (the first *letter*, even!).

To illustrate this a little further, it seems useful to turn to that different kind of default: literature. In his novel *The Secret Agent*, Joseph Conrad invites the reader to an Italian restaurant in London, which we see through the eyes of the assistant commissioner of police:

> The patrons of the place had lost in the frequentation of fraudulent cookery all their national and private characteristics. And this was strange, since the Italian restaurant is such a peculiarly British institution. But these people were as denationalized as the dishes set before them with every circumstance of unstamped respectability. Neither was their personality stamped in any way, professionally,

> socially, or racially. They seemed created for the Italian restaurant, unless the Italian restaurant had been perchance created for them. But that last hypothesis was unthinkable, since one could not place them anywhere outside those special establishments. One never met these enigmatical persons elsewhere. It was impossible to form a precise idea what occupations they followed by day and where they went to bed at night. And he himself had become unplaced. (1968, 149)

This is a beautiful example of the diluted globalism that precedes "inessential commonality" and of the latter's challenge to identity forged primarily through nationality. Here whateverbeing emerges inside that most generic of spaces, the Italian restaurant, which, despite Marinetti's attempts to the contrary in his *Futurist Cookbook*, bears only the most tokenistic stamp of its ethnic origins. Just as people are born in America every day with no other apparent purpose than to buy and wear Gap clothing, the citizens of the coming community emerge from the sheer success of capitalism, growing like poisonous mushrooms within the cracks of Empire. To Conrad's example we could add Vladimir Nabokov's *Bend Sinister*, which depicts a police state "where everybody is merely an anagram of everybody else" (in 1991, 353); just as Joan Riviere writes: "There is no such thing as a single human being, pure and simple, unmixed with other human beings . . . [for] other persons are in fact . . . parts of ourselves. . . . We are members of one another" (in Kilgour 1990, 240).

Instances continue to multiply, giving way to both claustrophobia and agoraphobia. In a confessional letter Keats admitted, "When I am in a room with People if I ever am free from speculating on creations of my own brain, then not myself goes home to myself: but the identity of every one in the room begins so to press upon me that I am in a very little time annihilated" (in Agamben 1999a, 113). Georges Bataille writes: "A man who finds himself among others is irritated because he does not know why he is not one of the others. In bed next to a girl he loves, he forgets that he does not know why he is himself instead of the body he touches. Without knowing it, he suffers from the mental darkness that keeps him from screaming that he himself is the girl who forgets his presence while shuddering in his arms" (1985, 6).

Margaret Atwood's *The Edible Woman* depicts a similar moment of threatening dissolution as the protagonist eats some cake and tea with her colleagues,

> their fluidity sustained somewhere within by bones, without by a carapace of clothing and makeup. What peculiar creatures they were; and the continual flux between the outside and the inside, taking things in, giving them out, chewing, words, potato chips, burps, grease, hair, babies, milk, excrement, cookies, vomit, coffee, tomato-juice, blood, tea, sweat, liquor, tears, and garbage. . . . For an instant she felt them, their identities, almost their substance, pass over her head like a wave. At some time she would be—or no, already she was like that too; she was one of them, her body the same, identical, merged with that other flesh that choked the air in the flowered room with its sweet organic scent; she felt suffocated by this thick sargosso-sea of femininity. She drew a deep breath, clenching her body and her mind back into her self like some tactile sea-creature withdrawing its tentacles; she wanted something solid, clear: a man; she wanted Peter in the room so that she could put her hand out and hold on to him to keep from being sucked down. Lucy had a gold bangle on one arm. Marian focussed her eyes on it, concentrating on it as though she was drawing its hard gold circle around herself, a fixed barrier between herself and that liquid amorphous other. (1969, 167)

If no woman is an island, it is possible to be drowned in a sea of femininity during a distinctly Freudian panic. Hence the protagonist's acquiescence to a masculine life buoy to shore up a fragile identity. Gender is obviously paramount in this example, and I have no wish to discount its importance. However, as with Deleuze's essences (discussed below), identity can be hermaphroditic, even after Zeus's vindictive surgery.

To get married, or otherwise co-habituate, is to enter into a more complex assemblage: that of the couple. Thus, we suddenly become a hyphenated being: Tim-and-Tess, Claire-and-Phil, George-and-Mildred, and even "Billary," "Bennifer" and "TomKat."[3] (Not to mention the fascinating overlaps between identical twins, which seem to take the ontic-origami of "the fold" to yet another level.) The self and the other become far more symbiotic.

No doubt the list of interpersonal interpenetrations is as inexhaustible as the people who record and inspire them. Together they reinforce the suspicion that our modern spin on identity is but an echo of the ancient attitude towards the same: Humanity represents only one part—and an unstable one at that—of the vast Cosmic Recycling Program.

On Essence, (or "The Immortal in Another Way")

> For a hundred years, (literary) madness has been thought to consist in Rimbaud's "Je est un autre": madness is an experience of depersonalization. For me as an

amorous subject, it is quite the contrary: it is becoming a subject, being unable to keep myself from doing so, which drives me mad. I am not someone else: that is what I realize with horror.

—ROLAND BARTHES, *A Lover's Discourse*

There are only three questions asked in art: who am I? And who are you? And what the fuck's going on?

—JOHN LANCHESTER, *The Debt to Pleasure*

I'm sitting next to a young woman on a crowded train. She starts complaining loudly to her gum-chewing friend sitting opposite. "I hate crowds," she says, as if, by some act of sheer will, she is somehow exempt from comprising part of it. By audibly articulating her displeasure, it seems she is attempting to invoke a spell to ward off an excess of otherness encroaching on her shopping bags and much vaunted "personal space." Mercifully, there are fewer people on my return journey. This time there are two young girls, about seven years old, captivated by a station billboard for a pair of women's shoes. The model stares through the window like an indifferent airbrushed Valkyrie bearing an emphatic but enigmatic message for every person who meets her unfocused stare. "She's beautiful," announces one of the girls, before adding, "I'm her." The other girl objects: "No you're not. *I'm* her." The first girl to speak—possibly a future leader in the burgeoning science of conflict resolution—finishes the matter by stating, "Okay. We're *both* her." Taken together, these two scenes capture more than just the mutual reinforcement of identification and advertising, for they speak of competing logics when confronted by alterity. In both cases the other is excessive: the first in terms of quantity, the second in terms of insistence or presence. And both cases show that the traditional existential anxieties expressed in the two epigraphs that introduce this chapter seep into the everyday language of adults and children alike.

For a number of years now, it has been something of a heresy, at least in the halls of "progressive" humanities departments, to suggest the existence of an essence. The concept itself, implying a stable property or quality which belongs to and defines an object, goes against the grain of the large share of post-structuralist theory which seeks to emphasize the "shifting, fluidic, Heraclitean flux" of postmodern existence. Certain hijackings of Lacan in particular have used subjectivity itself as the deconstruction site for the modern-liberal-enlightenment unit of the self, which has hitherto been viewed as stable, autonomous, sovereign, and ultimately atomic.

Ella Shohat and Robert Stam, for instance, lament the self-fulfilling–prophetic trajectory of so-called Western history "from Plato-to-NATO" (2002, 14). However, if we take the time to read the thinker named in the first of these terms carefully, a less coherent picture emerges. Indeed, we would be hard-pressed to find a more cogent portrayal of "multiplicity" than that provided by Diotima, the wise woman of Mantineia, speaking to Socrates in the *The Symposium*:

> The new is always left in the place of the old. For even in the same individual there is succession and not absolute unity: a man is called the same; but yet in the short interval which elapses between youth and age, and in which every animal is said to have life and identity, he is undergoing a perpetual process of loss and reparation—hair, flesh, bones, blood, and the whole body are always changing. And this is true not only of the body, but also of the soul, whose habits, tempers, opinions, desires, pleasures, pains, fears, never remain the same in any one of us, but are always coming and going. And what is yet more surprising is, that this is also true of knowledge; and not only does knowledge in general come and go, so that in this respect we are never the same; but particular knowledge also experiences a like change. For what is implied in the word "recollection," but the departure of knowledge, which is ever being forgotten, and is renewed and preserved by recollection, appearing to be the same although in reality new, according to that law of succession by which all mortal things are preserved, not by absolute sameness of existence, but by substitution, the old worn-out mortality leaving another new and similar one behind—unlike the immortal in this, which is always the same and not another? And in this way, Socrates, the mortal body, or mortal anything, partakes of immortality; but the immortal in another way. Marvel not then at the love which all men have of their offspring; for that universal love and interest is for the sake of immortality. (1999, 741–42)

Nevertheless, the bulk of readings of Descartes, Kant, and Hegel have produced a self with an essence, something unique to each person and located somewhere in either the mind, the heart, or the soul. In contrast to the B-movie histrionics of the "postmodern onion people"—who can shed layer upon layer without revealing a "true" kernel of selfhood—the more traditional, basic democratic unit (one monad, one vote) is the proud owner of an essence. Indeed, this latter notion has been so prevalent and persistent that I feel safe in saying that the reader need only check the current pop charts for a singer (alternatively, check the current talk shows for an expert guest) who advises the public to "be true to themselves." It is an individual's essence, unsullied by others, to which such people claim fidelity.

Beyond the metaphors of onions versus kernels, however, exist alternative conceptions of essence—Deleuze's in particular, freeing it from the prison of selfhood, just as Foucault freed the flesh from the prison of the soul. In contrast to those who want to throw out essence completely via the hostile charge of "essentialism," Deleuze's reading of Proust allows us to avoid making the mistake of throwing the bathwater out with the baby.

We are already familiar with the crucial passage in Proust, quoted in this book's introduction, concerning the gaggle of girls on Balbec's beach. This spectacular gang caused such a commotion within the internal vibrations of the narrator that he had trouble isolating one girl from the other. Their net effect created a lasting impression of an inchoate and gestalt arrangement of sheer presence, prior to the stamps of personality which are usually used to distinguish between them. But what is it that lurks *behind* such an apparition? What can we discover *beneath* the sands, which have become so unsettled?

It is impossible to adequately summarize Deleuze's delicate and complex presentation of "essence" in *Proust and Signs*, especially if we take into account the way his book playfully tempts the reader to see a distinctly non-Deleuzian transcendentalism or aestheticism at work in the ideas presented. But we must nevertheless attempt to harness a hazy conception of the way in which this particular essence circulates around and throughout our topic in order to "work through" the trauma of the second love. Indeed, Deleuze begins with the deceptively straightforward statement that "material meaning is nothing without an ideal essence that it incarnates" (2000, 13). Taking the exemplary case of the artist—specifically, Proust's attempts to salvage "the past as it is preserved in itself . . . [as] the instantaneous image of eternity"—Deleuze considers the vexed relations between memory, jealousy, and an artistic apprenticeship to signs constantly beaming contradictory messages from (almost) irreconcilable worlds. Thus, essences

> transcend the states of subjectivity no less than the properties of the object. It is the essence that constitutes the sign insofar as it is irreducible to the object emitting it; it is the essence that constitutes meaning insofar as it is irreducible to the subject apprehending it. It is the essence that is the last word of the apprenticeship or the final revelation. (38)

This "final revelation" amounts to a kind of meta–or anti–Archimedean point, in that essence "is not something seen but a kind of superior viewpoint" (110). Indeed, "the viewpoint remains superior to the person who assumes it or guarantees the identity of all those who attain it. *It is not individual, but on the contrary a principle of individuation*" (110, my emphasis).[4]

Until this essence is understood, however, "we always fall back into the trap of the object, into the snare of subjectivity" (38). Essence thus designates the unity of sign and meaning as revealed in the beloved or the work of art. But what does this really tell us? "What is an essence as revealed in the work of art? It is a difference, the absolute and ultimate Difference. Difference is what constitutes being, what makes us conceive being" (41).

Crucially (and this is what distinguishes Deleuze's schema from a transcendental or mystical concept), is his insistence that essence "does not exist outside the subject expressing it, but it is expressed as the essence not of the subject but of Being, or of the region of Being that is revealed to the subject" (43). Essence is "deeper than the subject," emitting signs *through* the subject to other subjects, although it cannot exist outside the subject. It is immanent to the system. Deleuze thus rephrases the existentialist maxim "Existence before essence" to something resembling "Existence *as* essence."[5]

Perhaps another example would be helpful at this point. Try to recall that moment when you look into your beloved's eyes and suddenly, *uncannily*, "see" a former love peering out at you. Or when you see a brother in the gestures of his sister, or the squint of a mother in her daughter. In these simplistic shortcuts we can understand how essence forges itself at the crux between difference and repetition.[6] Whether we have only two lovers, or two hundred (and we *always* have at least two, unless we happen to actually shack up exclusively with our mothers), then we begin to sense how "each love contributes its difference, which was already included in the preceding love, and all the differences are contained in a primordial image that we unceasingly reproduce at different levels and repeat as the intelligible law of all our loves" (68).[7]

Essence is thus the law of love's seriality, and the secret (often deceptive) signs that each successive lover presents to us, a secret that moves from assumed singularity to "an increasingly greater generality" (67). This is why, in Proust's words, "We shall need, with the next woman, the same morning walks" (in Deleuze, 2000, 69). Essence, therefore, acts as the fuel for an ultimately *machinic* relationship to relationships themselves. Suffering, jealousy, joy, confusion, doubt, and panic are no doubt experienced as unique and indeed lie at "the heart of a specific love." However, all these codified responses form a resonance machine within the pure potentiality, multiplicity, and contingency of amorous encounters.

Once again: Love is e(x)ternal. For Deleuze as for Proust, "nothing shows the externality of the choice better than the contingency that governs the identity of the beloved" (76). Resonating with Lola's nocturnal query "Why me?" the question of essence funnels back to the precariousness of subject-object overlaps and feedbacks: "[B]y what accident is it that Albertine incarnates essence when another girl might have done so just as well?" (76). Not only that, but who is this "I" who allegedly does the choosing? Who is it, exactly, who succumbs to "the automatism of love"?

> We shall not ask who chooses. Certainly no self, because we ourselves are chosen, because a certain self is chosen each time that "we" choose a person to love, a suffering to experience, and each time this self is no less surprised to live or to relive, and to answer the call, whatever the delay. (127–28)

Thus, and this really is the heart of the matter:

> It is not the subject that explains essence, rather it is essence that implicates, envelops, wraps itself up in the subject. Rather, in coiling round itself, it is essence that constitutes subjectivity. It is not the individuals who constitute the world, but the worlds enveloped, the essences that constitute the individuals. . . . Essence is not only individual, it *individualizes*. (43)

For Deleuze the emphasis on essence means that it is not purely a matter of infinite substitutability (since it is profoundly "irreplaceable"), but rather constant repetition. It is precisely at this point that we can see how it may be possible to reconcile the trauma of the second love, since the singularity of personal experience is rescued through several machinic motions: an increasing generality, an apprenticeship to signs, and ultimately a revelatory relationship with (lost) time. These machines combine to produce certain truths, via the work of art (which itself is a machine).[8]

Thus, love is exposed as a matter of hieroglyphs and hermaphrodites.

Proust himself states, "Imagination and thought can be splendid machines in themselves, but they can be inert; it is suffering that then sets them in motion" (in Deleuze 2000, 147). Part of this motion is the media that essence uses to manifest itself, since essence is comprised of "free substances that are expressed equally well through words, sounds, and colours" (47). Deleuze's valorization of the redemptive qualities of art seems more consistent with his own system when we remember that art and technology were only recently split by our own Zeus-like tendencies, and love itself may be the effort we bring to forging them back together. Paintbrushes, pianos, fountain pens, and laptop computers—these are what we use in place of Hephaestus's instruments to forge an encounter.

Indeed, according to media theorist Friedrich Kittler, love is primarily a matter of storage methods, so that art itself is figured as merely an ornate database of communication techniques. The information it stores (the lover's coding), however, always escapes the retrieval mechanisms of the day. He tells us:

> Only as long as the unchallenged and unrivaled medium of the book was able to simulate the storage of all possible data streams did love remain literature and literature love. . . . But a writer whose school teaches physics instead of philosophy objects. The combination of sensory data streams achieved by love is devoid of "permanence." It cannot be stored by any medium. Moreover, it loses "all individual character." That is, no real can pass through the filter of love. Which is why love does "not serve for the poet, for individual variety must be constantly present for him; he is compelled to use the sense sectors to their full extent," or, simply, to become a media technician among media technicians. . . . In other words: literature defects from erotics to stochastics, from red lips to white noise. (1999, 50–51, internal quotes from Rainer Maria Rilke)

As islands within the data stream, we feel our sense of individuality threatened by the global warming of "hot" media, such as the internet and other, less tangible networks. But this is no bad thing, as it leads to a recognition of whateverbeing and the junk coding both in our own DNA and in the lover's discourse itself.

So, just as Heidegger insists that the question concerning technology is nothing technological, we may suggest that "inessential commonality" is intrinsically a question of essence.

Divided and Entangled: The Coexistence of Coessences

> You will notice that I spoke of essence, just like Aristotle. So? That means that such old words are entirely usable.
>
> —JACQUES LACAN, *On Feminine Sexuality*

And so it seems that the term *essence* can be redeployed in ways which should not automatically make contemporary cultural critics—who usually sweat in Pavlovian fear at its mere mention—nervous. Essence thus becomes the context or precondition of/for mapping the inessential.

In order to elucidate this constantly elusive point, let us return to the question of technology in the form of communication. Bakhtin tells us: "The very being of man (both internal and external) is a *profound communication*. To *be* means to *communicate*" (1984, 287). And, no doubt, vice versa. But what exactly is being communicated if the inadequate binary model of transmitters and receivers is abandoned for something more nuanced and polycentric, something where point A and point B do not precede their encounter? *Communication* is then a loose term for the flow of information, the feedback loop, which precedes us as thinking subjects.[9] Just as Pierre Lévy argues that airplanes are the actual manifestations of a preexisting virtual network (1998)—in this case, of global travel—human beings as we know them can be considered as concrete cases of already established media vectors.

While it is impossible to deny that as so-called individuals we make up nodes in a social network, the implications become more significant when processed on the ontological level. Here the coherent, molar entity is brought into question at its very foundations, enabling us to get beyond the fatuous recognition of "influence," to the constitutional alienation of the self from its self. To borrow a concept from Benedict Anderson (1991), the self is then exposed as an "imagined community": population 1. The imagined aspect in this case is that the individual is in-dividual (as we saw earlier). Just as the nation is unthinkable outside the advent of technological developments such as print capitalism, the self is equally impossible outside the *exposure* of something felt to be intimate and private, but bequeathed from elsewhere, by anonymous—and perhaps even sinister—donors.

> If "communication" is for us, today, such an affair—in every sense of the word—if its theories are flourishing, if its technologies are being proliferated, if the "mediatization" of the "media" brings along with it an auto-communicational vertigo, if one plays around with the theme of the indistinctness between the "message" and the "medium" out of either a disenchanted or jubilant fascination, then it is because something is exposed or laid bare. In fact, [what is exposed] is the bare and "content"-less web of "communication." One could say it is the bare web of the *com-* (of the *telecom-*, said with an acknowledgment of its

> independence); that is, it is *our* web or "us" as web or network, an *us* that is reticulated and spread out, with its extension for an essence and its spacing for a structure. (Nancy 2000, 28)

In other words, "Being is communication. But it remains to be known what 'communication' *is*" (28).

If some readers would prefer this notion be presented in the more familiar mode of dialectical logic, then we can rephrase it through Hans Jonas:

> This dialectic is precisely the nature of life in its basic organic sense. Its closure as a functional whole within the individual organism is, at the same time, correlative openness toward the world; its very separateness entails the faculty of communication; its segregation from the whole is the condition of its integration with the whole. . . . Only complex functional systems afford the inner autonomy that is required for greater power of self-determination, together with greater variety of inner states responding to the determinations which impinge on it from without. . . . Only by being sensitive can life be active, only by being exposed can it be autonomous. And this in direct ratio: the more individuality is focused in a self, the wider is its periphery of communication with other things; the more isolated, the more related it is. (in Hansen 2000, n.p.)

As noted already in passing, it seems barely coincidental that one of the first philosophers to present himself to the world under the sign of both shame and defiance—Rousseau, via his epoch-shaping *Confessions*—was also tempted to roam the streets of Turin and flash his genitals at the washerwomen who gathered by the town's well at twilight.[10] These dubious crepuscular compulsions (flashing, writing, confessing) underscore the importance of exposure to the modern project of self-narration and self-consciousness. Hence there is a correlation between wishing to expose one's deepest secrets (thereby solidifying those "events," in Badiou's sense, which form the magnetic poles of autobiography) and the urge to expose the body which is increasingly seen to harbor such experiences, hoarding them jealously for this fledgling self, as a squirrel gathers nuts for winter nourishment.[11] On this theme Nancy notes, "We happen as the opening itself, the dangerous fault line of a rupture" (2000, xii).[12]

Such exposure is also evoked by the anxiety-producing imperative of love, which contains within it the shadows and ghosts of that more recent technology photography. That is to say, the fear of rejection or replacement is haunted by that special effect known as double exposure, where different times and places overlap in a single image, thereby collapsing the singularity and integrity of each moment. Thus, to say "I love you" and not have the sentiment returned is to be exposed on the narcissistic level of psychology, but also on the vertiginous level of ontology. This is so for the simple reason Sylvia Agacinski gives: "This inadequacy of sexual and mortal existences is already the inadequacy of bodies, which are less separate than it might seem and which are, before anything else, excrescences of other bodies" (1991, 16–17).

Agacinski qualifies this situation:

> All mammals' infants graft themselves to, or "plug themselves into," the body of their mothers while they breastfeed, such that each one is momentarily the organ of the other (and the infant is no less necessary to the breast than the breast to the infant). This troubling connection (one that troubles the opposition of the self and the other in the same way that sexual relations also trouble it) cannot be described as a subject/object relationship, any more than it allows one to say who or what is active or passive. (17)

In concert with Nancy, Agamben, and others, Agacinski believes that "the question *of the other* is only possible if the other is irreducibly plural, if it is *others*, and if they are not thought of in the perspective of an *us* (a collective subject)" (12). Moreover, she directly connects this insight with the "experience of what we call 'love,' an experience that cannot be thought of in terms of a 'bond' between two adequate concepts" (15). In other words, when we *fall* in love, we experience the vertiginous realization that existence is lived "in the mode of absence . . . entrusted to the other" (15). Thus, the ontological weakness of the subject is, at least for Agacinski, a precondition for falling in love (in contrast to the Cartesian or Kantian consciousness, which is free, autonomous, and capable of "resting upon itself").

In essence, then, love is figured as intimately and intrinsically *technical* and *communal*, since the weakness which nurtures its emergence is

> to *hold* (to remain, to remain "standing," to be, to be stable . . .) only to find supports or props "outside of the self": the earth, the mother, but also any form of support one can think of, including all the prostheses usually classed as technical objects or instruments. (Only God is supposed to rest upon himself absolutely.) (16)

This basically comes down to the understanding that if we are to avoid the well-meaning, circular superficiality of certain existentialisms—"Who is this I which exposes itself?"—it is necessary to ground the theory of exposure in a more sophisticated analytical framework. While several such are available, there are few more lucid accounts than Nancy's *Being Singular Plural*, which fleshes out Heidegger's *Mitsein* (with-being)—itself, according to Nancy himself, "still no more than a sketch" (2000, 44).[13] Nancy's book gives us a better sense of what is at stake in sharing singularities (or rather, the sharing *that is* singularity). Why, for instance, do people feel the urge to expose themselves, when we spend so much of our lives fearing exposure? Is the perverse thrill and/or relief of wallowing in controlled shame (exhibitionism) enough to account for that same pleasure in relief? Isn't something more subtle going on, challenging the very self-who-is-supposed-to-be-ashamed-via-exposure?

We often speak in very abstract terms, especially when discussing society at large. We utter sentences like "They say we should be eating less salt" or "They say miniskirts are back in." But who exactly is this "They"? The scientific community? Fashion designers? A loose conglomerate of experts? Well, the answer is both, all, and none, for "the they," as Heidegger calls them, are an empty category which functions as a particular horizon for ontic understanding. The *they* are the "subject" of everydayness and complicate the relationship of the I to the self. Rather than representing the mass of not-I's which form the anonymous backdrop for our own egocentric limelights, the they are "those from whom one mostly does not distinguish oneself, those among whom one is, too" (Heidegger 1996, 111).

For Heidegger, as for Nancy, the history of metaphysics has come up woefully short when attempting to think—or even acknowledge—the essential *with* that binds being together. *Being* in the singular does not precede being with other beings, they insist, but arises from this originary conjunction. "The world of Da-sein is *with-world*," writes Heidegger. "Being-in is *being-with* others" (112). As a necessary extension of this condition, "Knowing oneself is grounded in primordially understanding being-with" (116).[14]

Heidegger is at pains to delineate and encourage the authentic mode of being-one's-self—not, however, as a detachment of the self from the *they* (a breaking loose and declaration of independence) but, enigmatically, "*an existentiell modification of the they as an essential existential*" (122). It is around this point that many people's eyes glaze over, daunted by the sheer effort of keeping up with Heidegger's idiosyncratic vocabulary and his remarkably complex rendering of ontology. Indeed, if Nancy really believes *Mitsein* to be a sketch, we can only shudder in horror at the prospect of a more sophisticated portrait of proto-whateverbeing, perhaps springing from his own brow. However, Nancy's notion of being singular plural does indeed manage to map these concepts in a relatively intelligible way. Not only this, but he manages to counter Heidegger's "mistake" of beginning with *Dasein* before moving to *Mitsein* (and thereby avoids replicating the metaphysical hierarchy that Heidegger himself spent so much energy trying to undo).[15]

In presenting his theory of "this autistic multiplicity" (2000, xiii), Nancy sharpens the focus of our own discussion, in which humanity cannot be considered as something external to, and thereby superior to or in control of, information technology. Thus, "we *are* meaning in the sense that we are the element in which significations can be produced and circulate" (2, my emphasis). In simplified terms, the medium is the message, and the message is "us" (that is, "them"—since we are never coterminous with that self-narrativized construct known as "ourselves").

For Nancy the "plurality of beings is at the foundation of Being," and as a consequence, a "single being is a contradiction in terms" (12). For Agacinski, too: "The 'subject' only encounters the 'problem' of the other and of coexistence because it has begun by detaching itself (from the world and from others), and by forgetting that it is, before anything else, in-the-world and with-others. Nothing is more remarkable than this operation by which thought withdraws into itself and dis-engages itself from existence" (1991, 12). Because of this *originary* coexistence, meaning is immanent to its own circulation within the network. Without getting too involved in the question of "meaning," it is prudent to note that ontology and epistemology (that is, what we *are* and what we *know*) overlap through the primordial phenomenon of sharing. Thus, a recognition of plurality must occur before any attempt to understand the world, or else we are immediately sliding down Descartes's slippery slope toward atomism and ultimately even solipsism and psychosis.

Moreover, there is no "world" to comprehend in a holistic and representative sense—as Heidegger insisted in his discussion of the world picture[16]—simply because sense is produced in that transductive space enabling "the spacing and intertwining of so many worlds" (Nancy 2000, 5). What we refer to so casually as *the world* is not the aggregate of its citizens, both animate and inanimate, but the intersections of each locality. Against all the rules of grammar and the ontological principles they assume, the subject actually experiences Being in the paradoxical mode of the "first person plural"—a kind of indiscrete individuation. And as with our own primal scene of the first handshake, this spacing—this *between*—"does not lead from one to the other; it constitutes no connective tissue, no cement, no bridge" (5).[17] In fact, we should cease seeing ourselves as either connected or unconnected, for such increasingly digital options are too binaristic to account for the quantum valences of relations. As with Lacan, encounters are neither centrifugal or centripetal, but "the *inter*lacing of strands whose extremities remain separate even at the center of the knot" (1999, 5).

Of course, in our own lives, we do not necessarily think explicitly in these terms. And yet, as I have tried to show thus far and will endeavor to do so in the pages that follow, many minor epiphanies emerging from the cultural fabric attest to an affirmation of being singular plural (another mode, I suggest, of whateverbeing). For Nancy and other like-minded thinkers, the founding tension is between the radical substitutability of the subject and the unique event of singularity. Indeed, one could be forgiven for thinking that the villainous figure of the "individual" has been booed off the pantomime stage of contemporary Continental thought, only to be replaced by the garrulous hero of "singularity," a pale inversion of the universal figure it is supposed to usurp. Indeed, reading such material, it sometimes feels as if we find ourselves in a French-Italian remake of the *Life of Brian*: the assembled crowd prompted to admit, "Yes, we are all singularities" (followed by the single, desultory exception: "I'm not"). However, the difference between the liberal-humanist individual and the emerging figure(s) of singularity is ultimately decisive—especially considering the latter's acknowledgment of interpersonal interpenetration *preceding* any consciousness of selfhood. So, while we may still be tempted to say of an eccentric friend, "When they made X, they threw out the mold," we are now forced to realize that the crucial presence within such a statement is not so much X as the *they*.

Such far-from-trivial qualifications, as always, become more clearly defined when we approach them through the prism of love and technology. Since what is essentially under discussion is the question of community—as essence—these other two terms will allow us to grasp how they are also in fact generated by each other from the outset (which should not be thought of as a locatable point of origin, in either time or space, but rather as the pre-creation *always already*).

Nancy tells us that for Heidegger, "curiosity is the frantic activity of passing from being to being in an insatiable sort of way, without ever being able to stop and think" (2000, 19). Such curiosity is the passion driving the seducer, who mistakes the "uncircumspective tarrying alongside" of carnal knowledge with that more elusive form of acquaintance, the "authentic understanding of others." From this distinction it is easy to see how vulnerable Heidegger's notion of curiosity is to moralistic, monogamous interpretations (since the emphasis on "passion" does indeed point to a thoroughly subject-centered attitude toward the other—precisely as an "other" to be discovered, conquered, consumed, and so on).

For Nancy the question is not passion, but *compassion*, for this latter term recalls "the contagion, the contact of being with one another in this turmoil. Compassion is not altruism, nor is it identification; it is the disturbance of violent relatedness" (xiii). Thus, "[i]t is no accident that sexual curiosity is an exemplary figure of curiosity and is, in fact, more than just a figure of it" (20). Consequently, the erotic encounter is as much an opening up, an exposure, of Being itself as the mutual offering of two autonomous beings. The ontic discourse of sharing bodily fluids according to capitalist economics is thus itself exposed as a flimsy soap opera compared to the ontological experience of sharing a moment according to the circulation of meaning.

But why only a moment and not the eternity of the lover's discourse? Well, without delving into the admittedly constitutive temporal dimension of this point, the matter relates to the randomness of social life and the intrinsic use-by dates of events. These are due, no doubt, to human mortality, but also to boredom, compulsion, obligation, "drifting apart," and however else we wish to figure the figure of Time Itself. (As Badiou says, "Randomness, from which every truth is woven, is the subject's material" [1991, 28].)

As if anticipating my own discussion of Wong Kar-wai and Haruki Murakami, Nancy notes, "It is never the case that I have met Pierre or Marie per se, but I have met him or her in such and such a 'form,' in such and such a 'state,' in such and such a 'mood,' and so on" (2000, 8). Thus, "being in the mood for love" is more about the contingencies of context than about the contents of its characters:

> *Being singular plural* means the essence of Being is only as coessence. In turn, coessence, or *being-with* (being-with-many), designates the essence of the *co-*, or even more so, the *co-*(the *cum*) itself in the position or guise of an essence. In fact, coessentiality cannot consist in an assemblage of essences, where the essence of this assemblage as such remains to be determined. In relation to such an assemblage, the assembled essences would become [mere] accidents. Coessentiality signifies the essential sharing of essentiality, sharing in the guise of assembling, as it were. (30)

Deleuze states: "We wonder about what makes the individuality of an event: *a* life, *a* season, *a* wind, *a* battle, 5 o'clock. . . . We can call ecceities or hecceities these individuations that no longer constitute persons or 'egos.' And the question arises: Are we not such hecceities rather than 'egos'?" (1991, 95). Such rhetorical questions add weight to this general dismantling of the metaphysical building blocks of both romantic narrative and society itself, replacing the rationality of intention and will with the fundamental ventriloquism of our own thoughts. In this case, however, the puppet master is not God, nor even ideology or the Media, but the proverbial "next man" who could—in alternative circumstances or parallel universes—be "you."

Deleuze continues: "The imperatives and questions with which we are infused do not emanate from the I," since "another always thinks in me" (199–200). For those with a latent investment in the repressive apparatus required to maintain the illusion of homeostatic coherence within individual consciousness, this is a scary statement. But for those who, in contrast, believe that "the notion of subject has lost much of its interest on behalf of *pre-individual singularities and non-personal individuations*" (95), this is an intriguing and promising approach to questions of belonging.[18] Moreover, it becomes clearer why distressing "clinical" conditions, such as schizophrenia and multiple personality disorder, are reinflected as caricatures or excessive mutations of an ontological truth: a return of the repressed.

Let us take the example of Jimi Hendrix.

This historical figure is not simply the subject of a broader discussion—a now-deceased individual. When we talk of Jimi Hendrix, we are actually talking about "Jimi Hendrix": an assemblage of elements, including physiology (long fingers, left-handed coordination), "race" or ethnicity (diasporic African-American), technology (amplifiers, pickups, wahwah pedals, LSD), cultural events (Woodstock, Isle of Wight), plurality (audience, feedback), among a host of others. "Jimi Hendrix" is thus the quilting point for various discourse networks which have become emblematic of that cultural moment known simply as "the Sixties." Without belittling his undoubted and unprecedented talent, the term *genius* should be approached with caution, since it is saturated with the logic of the liberal-humanist "Man of Letters," an orientation which completely neglects the way brilliance or intelligence emerges dialogically from the clinamen, that is, from within a larger set of elements than the great Author or Artist himself (historically gendered as such). From this less familiar perspective, Jimi Hendrix is a spectacular node in the circuit (indeed, the node itself *as* circuit), himself part of a feedback effect as powerful and resonant as those between his Fender guitar and Marshall stack.

Similarly, we could point to the multitude of poets and musicians who talk of the strangely passive experience during the creation of "their" work. These testimonials are usually variations on the theme "The music was just flowing through me"—that is, from elsewhere. Art is less the fruit of a certain maturity and understanding than of the sparks given off by the hardwiring of the matrix itself (whether figured as langue, grammar, the noosphere, the collective unconscious, or whatever). The human element is most certainly an operator within this assemblage, with a certain amount of choice at his or her disposal ("I think I will use the blue paint rather than the orange"), but as any Freudian, ontologist, or systems theorist could tell us, our choices are very rarely "our own," pure and simple, exercised outside invisible psychosocial pressures.[19] (And here we see shades of Adam Smith's invisible hand, which moved from the external sphere of macroeconomics to the internal sphere of libidinal economies in the early twentieth century.)

As always, *those three little words* are never far from our discussion. Since for Nancy and others, "[p]resence is impossible except as copresence" (2000, 62), then

> the relation as such is nothing other than its own representation, the symbolic is what is real in such a relation. By no means, however, is such a relationship the representation of something that is real . . . but the relation is, and is nothing other than, what is real in the representation—its effectiveness and its efficacy. (58)

The paradigm for this is, interchangeably, the utterance "I love you" and the enormous, jealous, ambivalent chain-love-letter addressed to ourselves known as the spectacle.[20]

Nancy asserts: "There is no society without the spectacle because society is the spectacle of itself," just as "the truth of the play of mirrors must be understood as the truth of the 'with.' In this sense, 'society' is 'spectacular'" (67–68). Love simultaneously traces the productive "limits" of incommensurability (considered as empirical boundaries only by the romantic trope of unity, fusion, transcendence, and so on) and "Being-together . . . defined by being-together-at-the-spectacle" (51).

The vacuum of democracy rests on the same premise: a universal subject without-qualities, and therefore without any political power or ontological clout whatsoever.[21] For Nietzsche this empty multiplicity is far from pleasant: "What is disagreeable and offends my modesty is that at bottom I am every name in history . . . I go everywhere in my student's coat, and here and there slap somebody on the shoulder and say, Siamo contenti? *Son dio ho fatto questa caricatura*" (in Durham 1998, 222). Yet, in *Ecco Homo*, Nietzsche's last "finished" manuscript, he passionately embraces the fact that he is every name in history. In a Dionysian precursor of Joyce's H.C.E. (Here Comes Everybody), Nietzsche seizes on the potential power of whateverbeing. For as soon as we accept the essential nature of being singular plural, we cannot help but take that next step and realize that "what will not be built any more henceforth, and cannot be built any more, is—a society in the old sense of that world. . . . *All of us are no longer material for a society*" (in Durham 1998, 220).

In Essential Commonality

> We approach a society/ Without a society
>
> —JOHN BROCKMAN, *Afterwords*

> But we do not even say we . . . a we that would be without content.
>
> —MARCEL PROUST, *Remembrance of Things Past*

Hopefully, we are now equipped to finally consider Agamben's notion of the coming community, especially his seductive call for an "inessential commonality." In the last section we broached the concept of an essence, a belated recognition or unifying principle (an "immortality in another way") which exists *through* the individual but does not *belong* to any particular individual. But how does such a notion differ from, say, the Marines? Here, in principle at least, all identity is gradually extinguished through discipline and abusive conditioning, so that the soldier behaves like an unthinking cog in the military machine (as depicted in Kubrick's *Full Metal Jacket*). Can we say that the soldier, like the character actor, is an exemplar of whateverbeing?[22]

Before we answer Yes, we must consider certain qualifications. In this instance, whateverbeing is harnessed by forces which pervert the potential of the concept itself. In Deleuzian terms, the soldier is a highly territorialized form of whateverbeing. But just as the war machine always threatens to break free from the harness of the state, as a wild horse struggles to throw off its rider, whateverbeing will always exceed the drills and procedures of the military mandate. Whateverbeing consistently deterritorializes any organized attempt to defend or attack actual territory, which, of course, is the traditional directive of the soldier.

The soldier has a sibling: the mystic disciple, who seeks to lose all distinguishing qualities of self in order to dissolve in the unity of the godhead. This man or woman without content, without ego, is the mirror image of the soldier, for both are, from a certain angle, possible subjects of a coming community. Both act according to a profound perception of what it means to *be* an essence rather than *have* an essence. And both point to the possibility of a world where newborns are not even given proper names, that marker of selfhood on which we hang our evanescent identities like on a meat hook.

But it seems that these qualifications lead to disqualification, for we cannot, in good faith, list the soldier or mystic alongside Agamben's exemplars of whateverbeing—tricksters, fakes, assistants, and 'toons.[23] Certainly, the soldier and mystic are infinitely substitutable, but they dwell in the place of the Same, not the Other. They try to extinguish or neutralize difference. And ultimately, this difference is decisive. The marine and the mystic each belongs to a common body: the corps and the corpus, respectively. As such, they represent part of a transcendental category and thus exclude themselves as exemplars of whateverbeing. As opposed to Deleuze's cosmology, "something"—in this case the institution of the military or religion—precedes individuation and exists as a supplement to the aggregate of individuals. In the coming community, essence cannot be conceived outside of its own incarnation(s).

To the mystic and the marine, the loss of self within a greater essence can be a form of *communion*—something anathema to contemporary critiques of community. For them this loss of ego is still a utopian prospect. To the liberal-humanist, in contrast, this scenario is a dystopian vision of pseudo-subjects, with only dronelike status in the service of a faceless state, tyrannical leader, or delusional opiate. How could love flourish in such a situation? they ask. The simple answer is that it cannot, which is no reason to reject the scenario outright. We must seriously consider the suggestion that love (or at least the assumptions of the lover's discourse—the conditions which produce it in its current form) acts as a moralistic barrier *against* our struggle with the nationalisms and factional religions which currently and consistently wreak havoc on this world.

According to its current coding, love allows a person to recognize, and thus treasure, the essence of another person. Whateverbeing would find this notion nonsensical, for the reasons detailed thus far. Love exists in the barracks or the ashram only when it is channeled toward a more abstract principle (the mother country for the soldier, and the godhead for the mystic). This is still a dangerous situation, due to the co-option by forces which attempt to harness such love and redirect it into indignance, dogma, and, finally, its supposed opposite, violent hatred. The love of one's own country, family, or god translates too easily into the hatred of another's country, family, or god to be considered a different emotion or orientation. As Freud, Girard, and countless others could tell you, hate is immanent to love.

These movements trace the thin red line of Terence Malick's majestic movie of the same name (1998), a virtual line which separates sanity from madness, wartime from peace, life from death, love from hatred, you from me, and the metaphysical from the political. This line is so easily crossed that its status as something which separates is called into question, and it becomes rather a "line of flight" from the solidifying forces which seek to impose the will of the war machine. Just as essence itself does not exist outside those who make it manifest, Eros and Thanatos do not stalk the planet in any other form than the population of the day, a population animated by different cultural and historical epochs and agendas.

In *The Thin Red Line* (1998), animals and indigenous people continue with the work of the cosmos, indifferent to the European enactment of the same that is unfolding around them. This is not in order to romanticize a more "organic" connection with the world that the European has lost, but rather to highlight the wider patterns which continue both beyond and within modernity, challenging the teleological rationalizations for homeopathic genocide. Indeed, the "natural world" features highly in all three of Malick's films, suggesting a cosmic connection continuously undercut by the disorientations of *Dasein*. (Before making films, Malick translated Heidegger's *The Essence of Reasons* for Northwestern University Press.)[24]

Private Bob Witt (Jim Caviezel) grapples not only with World War II itself but with the conflict between gestalt philosophies of humanity and the implications of an inessential commonality. We sometimes see him fighting and then trying to escape, striving to leave the frame of the film itself and join Thomas Pynchon's Slothrop in pure diffusion.[25] And as a pursued hybrid of mystic and soldier, Witt questions the limitations of both. Sergeant Eddie Walsh (Sean Penn) is the world-weary guardian of human dignity, outpaced by the repercussions of such so-called civilization. At one point the voice-over which punctuates the film ponders on a vulgar version of Deleuze's essences: "Maybe all men got one big soul where everybody's a part of. All faces are the same man, one big self. Everyone looking for salvation by himself. Each like a coal drawn from the fire."

The writings of Blanchot are particularly sensitive to these repressed affinities and can serve as a second screening for *The Thin Red Line*, pointing to the place "where an episodic community takes shape between two beings who are made or who are not made for each other." Thus, "a war machine is set up or, to say it more clearly, the possibility of a disaster carrying within itself, be it in infinitesimal doses, the menace of universal annihilation" (1988, 48). Such a perspective props up both the cinematic event of Malick's meditation on violence and the sacred as well as the insistent cliché "All is fair in love and war."

To do justice to the other (including the other of ourselves) means to keep the space of subjectivity open. It is to be "whatever," which Agamben tells us is not simply a matter of indifference, nor of "It does not matter which," but also etymologically contains the opposite, "being such that it always matters" (1993a, 1).[26] This involves ethically negotiating (or, further still, metaphysically exposing oneself to) alterity, a procedure explicitly explored in the chapter on "Ease" and the Talmud in *The Coming Community*:

> At the point when one reaches one's final state and fulfills one's own destiny, one finds oneself for that very reason in the place of the neighbor. What is most proper to every creature is thus its substitutability, its being in any case in the place of the other. . . . This substitution no longer knows a place of its own, but the taking-place of every single being is always already in common—an empty space offered to the one, irrevocable hospitality. . . . In this community there is no place that is not vicarious. (23–24)

Thus:

> Against the hypocritical fiction of the unsubstitutability of the individual, which in our culture serves only to guarantee its universal representability . . . [the coming community] presents an unconditioned substitutability, without either representation or possible description—an absolutely unrepresentable community. (24–25)

Agamben denotes this unrepresentable space as "Ease," noting, for our purposes, that the Provençal poets also used this word as a technical term "designating the very space of love. Or better, it designates not so much the place of love, but rather love as the experience of taking-place in a whatever singularity" (25).

The coming community figures belonging *as* becoming (be-coming) and thus drives a wedge between familiarity and belonging. (A techtonic analysis would insist that it is technology which renders this general stream of becoming, allowing the reconceptualization of being in relation to culture-saturated time.) As Agamben puts it:

> The Whatever in question here relates to the singularity not in its indifference with respect to a common property (to a concept, for example: being red, being French, being Muslim), but only in its being *such as it is*. Singularity is thus freed from the false dilemma that obliges knowledge to choose between the ineffability of the individual and the intelligibility of the universal. (1)

What has historically been presented in terms of "the problem of individuation" is now only a pseudo-problem. Agamben's inessential commonality partakes of "a solidarity that in no way concerns an essence. *Taking-place, the communication of singularities in the attribute of extension, does not unite them in essence, but scatters them in existence*" (18–19). But as we saw in the previous chapter, it very much depends on how you define essence. Indeed, it seems almost impossible, at least with our current conceptual tools, to map the movements of love, technology, and community, under the sign of whateverbeing, *without* recourse to the kind of essence as described by Deleuze. This forces me to make the perhaps bold claim that what Agamben calls "inessential commonality," Deleuze simply calls "essence." This is not to counter or confuse Agamben's notion but to unpack it further, although perhaps in a different direction from that which he intended.

As we have seen, whateverbeing is "neither particular nor universal, neither one nor multiple" (Agamben 1993a, 17). It resides in pure potentiality—a potentiality even after the (f)act—and thus speaks of an "incessant emergence" (20, 56). We can see this process occurring in the human face, which "is neither the individuation of a generic *facies* nor the universalization of singular traits: It is whatever face, in which what belongs to common nature and what is proper are absolutely indifferent" (19).[27] The enigma lies in a movement we have been taught to take for granted but that in fact describes a two-way transition that destabilizes the categories which shore up the representation of conceptuality itself. These include the movement from potentiality to act, language to word, common to proper, form to example, and vice versa.

Phrased differently, Agamben asks if there is such a thing as the letter *p* outside all those idiosyncratic inscriptions of it made by different human hands. Accordingly, the human face is not so much a variation on a theme, but a theme on variation. For our purposes it is sufficient to note that these two examples—writing and what Deleuze calls "faciality"—are technologies. Moreover, these technologies combine at the point of the interface, or the "interfaciality," of our contemporary networked systems (from which we ourselves incessantly emerge, as we saw in chapter 4). Hence, Agamben claims: "Whatever is the thing with all its properties, none of which, however, constitutes difference. In-difference with respect to properties is what individuates and disseminates singularities, makes them lovable" (19). Whenever we posit the technological, we also evoke the amorous and erotic: "The singularity exposed as such is whatever you want, that is, lovable" (2). In other words, whateverbeing is inconceivable without desire. The tricky aspect (only one among many questions which Agamben's book leaves to its readers) is the way in which desire circulates in the "free *use of the self* . . . that does not . . . treat existence as a property" (28–29). Simply put, how would the libidinal economy of the coming community differ from the one we live with, but perhaps do not understand, today?

In his major contribution to the debate, *The Inoperative Community*, Jean-Luc Nancy preempts Agamben's critical position, stating that thinking "community as essence" is in effect "the closure of the political." He continues in terms which should now be familiar to us, stating that community resists "letting itself be absorbed into a common substance" and, further, that "Being *in* common has nothing to do with communion, with fusion into a body, into a unique and ultimate identity that would no longer be exposed" (xxxviii). Nancy thus poses his key question: "How can we be receptive to the *meaning* of our multiple, dispersed, mortally fragmented existences, which nonetheless only make sense by existing in common?"—a commonality that does not stem from "the will to realize an essence" (1991, xl).

According to Nancy's system, the individual is "merely the residue of the experience of the dissolution of community" (1) and thus bears witness to the unraveling of "the autarchy of absolute immanence" (4). As a consequence, community is "neither a work to be produced, nor a lost communion" but rather a "space itself, and the spacing of the experience of the outside, of the outside-of-self" (19). After establishing this conceptual framework, Nancy emphasizes the importance of love, since love "exposes . . . the incessant *incompletion* of community. It exposes community *at its limit*" (38). Our lives are therefore figured as "an infinite migration through the other" (90), with the result that the refrain "I love you" becomes a declaration where "I" is posed only by being exposed to "you" (89). The self does not precede the other, just as the heart does not precede heartbreak (99).[28]

Hence the crucial paradox that "Love in its *singularity*, when it is grasped absolutely, is itself perhaps nothing but the indefinite abundance of all possible loves, and an abandonment to their dissemination, indeed to the disorder of these explosions" (83, my emphasis). But is it possible to resolve the tension between whateverbeing and singularity, since these concepts seem to tug at each other's frayed sleeves? Before answering, we should note that being in common is not to be confused with a "common being," the sympathetic Ur-structure of liberal-humanism. Thus, we cannot begin to repose the question of community without conceding the crucial role love plays in discourses of belonging. For as Agamben notes, "Love . . . conditions precisely the possibility of knowledge and the access to truth" (1999b, 186).

Whateverbeing: A Reprise

> Lovers go to the limit of the improper in a mad and demonic promiscuity; they dwell in carnality and amorous discourse, in forever-new regions of impropriety and facticity, to the point of revealing their essential abyss. Human beings do not originally dwell in the proper; yet they do not (according to the facile suggestion of contemporary nihilism) inhabit the improper and the ungrounded. Rather, *human beings are those who fall properly in love with the improper, who—unique among living beings—are capable of their own incapacity.*
>
> —GIORGIO AGAMBEN, "The Passion of Facticity"

> What is astonishing is not that something was able to be, but that it was able to not not-be.
>
> —GIORGIO AGAMBEN, *The Coming Community*

Rimbaud is constantly exhumed in order to reiterate his trademark phrase, "Je est un autre" (I is another). It takes the meticulous ventriloquism of Agamben, however, to remind us of the very next sentence which follows this modern mantra: "If brass wakes up a trumpet, it's not its fault" (1999b, 114). Only when the two points are made together does the significance of the utterance really emerge. It not only complicates all the epistemological distinctions between form and content but wakes our own slumbering maestro; the very same who conducts us through the comfortable confusions that make up our daily lives. For if we replace "brass" here with the term "essence," and "a trumpet" with "existence," then the homology moves closer to home.

Employing the kind of paradoxes which enrage those who dislike deconstruction (and giving only headaches to those more open to this project), Agamben states: "Whatever is singularity plus an empty space. . . . *Whatever, in this sense, is the event of an outside*. . . . [It] is, therefore, what is most difficult to think: the absolutely non-thing experience of a pure exteriority" (67). The coming community would thus constitute "the experience of being-*within* an *outside*" (68), since whateverbeing "is not an essence that determines an existence, but it finds its essence in its own being-thus, in its being its own determination" (93). The aforementioned exteriority of essence is what gives Agamben the theoretical leverage to keep this loaded term within his own system. "Only in this sense," he writes, "can we say that essence envelops—*involvit*—existence" (98). So we see that despite appearances, essence *does indeed* have a role to play in the figure of inessential commonality.

Further evidence can be found in Heidegger's influential thoughts on the matter, as summarized in these two sentences from his "Letter on 'Humanism'":

> To embrace a "thing" or a "person" in its essence means to love it [*sie lieben*], to favor it [*sie mögen*]. Thought in a more originary way, such favoring [*mögen*] means to bestow essence as a gift. Such favoring is the proper essence of enabling [*Vermögen*], which not only can achieve this or that but also can let something essentially unfold [*wesen*] in its provenance, that is, let it be. (in Agamben 1999b, 199)

For Heidegger the passions make up "the basic modes that constitute *Dasein* . . . the way man confronts the *Da*, the openness and concealment of beings, in which he stands" (198). Unlike anger, joy, and other "affects," love and hate—as passions—"traverse our Being from the beginning."

Speaking for his mentor, Agamben states that passion represents "the most radical experience of possibility at issue in *Dasein*: a capacity that is capable not only of *potentiality* (the manners of Being that are in fact possible) but also, and above all, of *impotentiality*" (1999b, 201). The Heideggerean strategy of "radical passivity," of actively experiencing the passivity of Being, thus becomes the provisional answer to Agamben's own question: "Where, and how, can a subject be introduced into the biological flow?" (1999a, 124–25).

Since the subject can be posited only through language and in relation to the other, we are left with the pure shifter "I" (116): "Indeed, 'I' signifies precisely the irreducible disjunction between vital functions and inner history, between the living being's becoming a speaking being and the speaking being's sensation of itself as living" (125). Anyone blessed with even the most minimal share of reflexivity as a speaking subject would acknowledge that "the fragile text of consciousness incessantly crumbles and erases itself," and thus we must come to terms with "the constitutive desubjectification in every subjectification" (123).

As David Odell pointed out much earlier, "intersubjectivity" seems like an inadequate concept for this radical potential at the heart of substitutability. ("Why me and not one of those other girls?" asks Lola. In contrast, a young Italian man randomly selected for execution by Nazi officers, in the writings of Antelme, refuses to speculate on the question "Why me, instead of someone else?")[29] The space between the killer and the killed—or indeed between the witness and the victim—produces the shame of singularity, while simultaneously fostering the horrific conditions for access to essence. Intimacy is the name Agamben gives to "a proximity that also remains distant, to a promiscuity that never becomes identity" (125). Or in Odell's words:

> Myself as a being amongst beings is the lie through which the truth of Being can first be glimpsed. The lie is known to be a lie because it cannot account for the difference that grounds the 'amongst,' and so it points inexorably to the non-place of Being where it is the very non-existence of difference which grounds indefinite plurality. This is strange because in its own realm Being is utterly undifferentiated and yet in expression, or phenomenality, it is what makes distinction possible, like a tiny mote in a super-saturated solution that triggers crystallisation. And in this manifestation that mote is found at the point where the crystallisation is densest and hence most fractured, in the centre of what is called "I." (2001)

That is, the time has come to drain the self of those narcissistic essential oils which we constantly use to massage our solitary egos. And hence the force of the task ahead: not to see others simply as other-mes, or even me as other I's, but the ex-static, essential flux of . . . well . . . of *what*, exactly? Here we fall back on corrupt terms, the most obvious being *God*. Indeed, it was this frustration which fueled most of Nietzsche's driven scribblings, and even he resorted to old, old words, such as *Dionysus*. We suffer from the eternal frustration of searching for a word both on the tip of the tongue and in the back of the brain, teasing us with a bittersweet foretaste. Is it a word we've forgotten, or a word we've yet to coin for a creature we've yet to name?

Hence the difficulty in avoiding this hoary old chestnut: "Who is this 'I' asking, 'Who is this "I"'?" Is it the "I" of grammatical postulation? The "I" of psychological narcissism? The "I" who speaks neutrally in public, who is indeed feeling fine and encouraged to continue having a good day, or the "I" who can't sleep; the one who Levinas stated was no *I* at all—simply a frazzled kind of empty insomnia?

We are familiar by now with the various post-structuralist concepts of the subject which displace or disperse the self into a constant flux or multiple network, according to which, the self is least of all where we think it is—that is, when we say or think the I-word. But in contrast to concepts which insist there is no center to the subject, perhaps we should entertain the notion that there *is indeed* an organizing principle (or, yes, even center), which *changes* from moment to moment. The self is thus akin to the eye of an electrical storm, which flickers and moves according to both centrifugal and centripetal pressures, a structuring absence of enforced calm. The "I" which inhabits meditation is very different from the "I" of orgasm, just as the "I" standing before an employee is inconsistent with the "I" standing before a magistrate. This is as obvious as it is opaque, context-specific (or "situational") time ushering the subject into a semicoherent shape. Heidegger of course soaked his lederhosen with sweat in his monumental attempt to convey this idea to us.

What we must now confront is the next question: "Who is this 'we' *refusing* to ask, 'Who is this "we"'?"

SEVEN

Mind the Gap

> For there is no such thing as a man who, solely of himself, is only man.
>
> —MARTIN HEIDEGGER, "The Question Concerning Technology"

> Who I am for you and who I am for me is not the same, and such a gap cannot be overcome.
>
> —LUCE IRIGARAY, *To Be Two*

At Waterloo underground station in London, all commuters are warned repeatedly by loudspeakers to "mind the gap" when a train approaches the station, a reference to the rather large space between the platform and the carriage. This piece of advice is just as useful when considering the various gaps between the self and the other, as when stepping from the platform onto the tube.

Take, for instance, this extract from Jean-Paul Sartre's *Being and Nothingness*:

> I experience myself as any transcendence: to go from the subway station at "Tracadéro" to "Sèvres-Babylon," "They" change at "La Motte-Picquet." This change is foreseen, indicated on maps, etc.; if I change routes at La Motte Picquet, I am the "They" who change. To be sure, I differentiate myself by each use of the subway as much by the individual upsurge of my being as by the distant ends which I pursue. But these final ends are only on the horizon of my act. My immediate ends are the ends of the "They," and I apprehend myself as inter-

> changeable with any one of my neighbours. In this sense we lose our individuality, for the project which we are is precisely the project which others are. In this subway corridor there is only one and the same project. (1956, 424)

As fate would have it, this particular quote of Sartre's is also used by Luce Irigaray in her book *To Be Two* (2000) as an example of what she believes is the misguided and masculinist method of approaching intersubjectivity. Instead of this rather traditional metaphysical paradigm, Irigaray offers her own alternative—"a new philosophy of the caress" ultimately leading to a form of "in-stasy." Unfortunately, for the most part, Irigaray's response is a disappointing devolution into sanctimonious Hallmark-card-meets-Helen-Reddy existentialism.

Irigaray states, "Humanity reaches fulfillment between the two genders" (33). So far, so bad. For while lesbians will no doubt be glad to hear that "[w]e are not complementary or supplementarity to each other," they may be less thrilled with this claim: "To be a woman necessarily involves—as far as human essence and existence are concerned—to be in relationship with man, at least ontologically" (34).

No doubt it would be churlish to quarrel with Irigaray's attempt to "propose relationships between two which are more human, more pleasing" (37), and it is indeed tempting to applaud her realization that "I can be a bridge for you, as you can be one for me. [But] [t]his bridge can never become the property of either" (43). However, these metaphysical gains compromise themselves as soon as she starts blaming that poor, misunderstood category "artifice" for annihilating authentic relationships between subjects. (As if "art" is not an essential part of being in love. Indeed, one wonders what Irigaray intends to *do* with her beloved, without such examples of artifice as music, cooking, candles—even conversation.)

Thus, for Irigaray, "linking myself to the other protects me from the alienation and fascination produced by the fabricated, 'manufactured' world which surrounds us, preventing a communal between-us" (38).[1] Her polemic goes on to argue that, in our increasingly instrumentalized society, "sexuality is left uncultivated, or rather, assimilated to a techne which does not take intersubjectivity into account" (40). However, as we have exhaustively detailed over the course of this book, *techne is* intersubjectivity—and then some. (This is why when Irigaray denounces "the latest Western philosophers," she conveniently avoids engaging with Nancy, Agamben, Agacinski, Lacan, Deleuze, Foucault, et al.—as if philosophy ended with Sartre.)

We must therefore be wary of arguments based on the suggestion that "we" have let technology get "out of control," since this assumes that we can confidently trace a line between the human and the nonhuman, the natural and the artificial. Moreover, this assumption leads to that commonly mistaken approach whereby love is figured as being the antidote to technology—as if Eros and Aphrodite were sleeping in a mountain somewhere, waiting to emerge and slay the bastard children of Hephaestus and Mars. Such arguments assume that "we" were once masters of *techne* and knew how to harmonize it with our lives, a knowledge or capacity we have since lost. As a consequence, these perspectives ignore the immanent logic within technology, which seduces and produces people as much as it forms the conditions of its own critique. (How could we denounce technology without language, which is itself a technology?)

Technology is not a machine, or the aggregate of all machines, which can somehow be turned off or dumped into scrap-metal yards when we become tired of it. Even now as I write, scientists are frantically trying to out-clone each other, far ahead of any legislation which would seek—mostly in vain—to control such practices ("in vain" because they are not practices occurring in a vacuum but the manifestation of desires forged by the epoch itself). This is not to argue for an immense techno-determinism, of which we are but an element in the assemblage. Or rather, it *is* to argue this but with the proviso that we recognize the potential (*puissance*) which comes from being a *privileged* element—that is, an operating element that reflects technology and is reflected within it, *through* it. (Perhaps if I used the word "discourse" or "culture" rather than "technology," skeptics would breathe a little easier.)

Irigaray looks forward to a time when we slow down the pace of our rampaging technoculture and return to "picking blackberries and doing pirouettes on the grass." This sounds all very pleasant but is hardly a useful response to the fabricated world which terrifies her so. After all, the transcendent salvation she seeks occurs under the sign of love, and isn't love something we *make*? Isn't love a fabrication—"Let's make love"—a human construction as real as our other constructions? In love we "bring something forth," to evoke Heidegger. Love is thus a *techne* and should not be considered as fundamentally antagonistic to the airplanes which bring lovers closer as much as whisk them away (a reference to Irigaray's resentment toward the public world, which demands so much time of her unnamed lover).

Herein also lies the limitation of Irigaray's project—that is, the complete denial of the many. Her utopia is always "to be two" and thus suffers from the suffocation of all binaries. In her most recent work, Irigaray shows her hand: specifically, a fistful of tarot cards, all marked with the ideal, happy couple, existing in a New Age no-place. This couple appears apolitical, heterocentric, and allegorical—and, most significantly, protective to the point of paranoia, for Irigaray seeks to

> drive away what intervenes between us. . . . To push to the outside of our relationship everything which disturbs, quantifies and compares. To repel whoever envies, mimes and wishes to appropriate. To distance whoever pretends to be or to rob me of who I am, you of who you are, us of who we are. To collect ourselves until we can escape all of these dangers. (15)

Such dangers include the television, the radio, and even neighbors. Irigaray buys a house in the country so she can be alone—completely alone—in a godless, ribless Eden with her neo-Adam.

No wonder her lover flees at every opportunity. No wonder he sounds tired and stressed on the much-hated answering machine. Irigaray blames the outside world for his exhaustion, but perhaps it is more the case that he finds it hard to live up to a love figured as outside all social biography. Who are these two? What happens if he leaves her, or her him? Suddenly, all those irrelevant, covetous mimics are *people* again, not just the background white noise of humanity on which Irigaray paints her Far Eastern characters. (But then again, as a man myself, I am predisposed to prefer "the relationship between the one and the many, between the I-masculine subject and others: people, society, understood as *them* and not as *you* [17]." Aren't I?)

Identity is the key issue for Irigaray, and it must be protected at all costs, even from her lover. But at what point can she confidently say, "This is me," and not the echo of the other(s)? Irigaray seems to resent the with-being of being-with, beyond the proto-familial dyad.

In attempting to safeguard her complete and essentially gendered identity, Irigaray is trapped in the dialectics of intersubjectivity, as much as she attempts to sidestep the appropriating gesture of synthesis.[2] "What makes me one, and perhaps even unique, is the fact that you are and I am not you"

(16), a comment which turns back on itself, in a meta-narcissistic loop. "To be two" is thus the imperative of a partitioned human essence, and Irigaray's gender is thus "irreducibly feminine."

If I harp on about the dubious distinction between *humans* and *technology*—with an almost fuguelike persistence—then it is only in direct ratio to the resistance most people put up to recognizing the essential symbiosis between these two terms. I know this because my students are willing to go to enormous lengths to prove their potential autonomy from the evil clutches of technology, not only renouncing their cars, houses, mobile phones, and makeup, but also their clothing, fire, and even language. Even so, they usually miss the irony that in going to such lengths to prove they are sovereign humans, my students have in fact hypothetically reverted to the kind of "animal" existence which has traditionally been considered that from which Man distanced and distinguished Himself. What Agamben calls the "anthropological machine" is fueled by the discourse which says, "Humans are those animals which do not (i.e., no longer) live like animals." We are the exception, since we are the creatures that have successfully harnessed—or been harnessed by—technology.

"The human" is thus coterminous with the "posthuman," since both are created with the same stroke. As soon as the simian picks up the bone, we are on the slippery slope to space travel, and thus "we" are cyborgs at the origin. As Bernard Stiegler and countless others have insisted, the posthuman has nothing to do with chronology. It is not a matter of animal—human—posthuman, but rather a leap from animal to posthuman, simultaneously incorporating and bypassing "the human."[3] As soon as we become technical, prosthetic creatures (that is, as soon as we speak, wear clothes, make tools, and so on), then we are post/humans. We are *always already* within that enframing mode of being that relates to the question concerning technology. Thus, to be already after a certain kind of humanness—even at the origin (or the "default of origin")—is to constantly negotiate, test, and redefine what the human is, has been, and may be.[4]

And yet the resistance continues with a remarkably stubborn tenacity. Humans simply don't want to give up their self-assigned precious place in the modern cosmological hierarchy. Those definitions of technology which expel this phenomenon outside of the human sphere, quarantining it in "objects" and "machines" and "artificial entities," do so according to the logic of apartheid. There is a fundamental conflict of interest at work in such schemes, since those who typologize unconsciously know what is at stake in that very process (that is, the mutating legacy of the romantic-humanist myth). To paraphrase Bourdieu, then: "Who will classify the classifiers?"

The Möbius Striptease

> Machines and humans are basically incompatible.
>
> —NICOLS FOX, *Against the Machine*

> We cannot, of course, reject today's technological world as devil's work, nor may we destroy it—assuming it does not destroy itself.
>
> —MARTIN HEIDEGGER, "The Principle of Identity"

What does it mean to be a neo-Luddite at the beginning of the twenty-first century? No doubt, many of us are familiar with the urge to throw our computer out of the window or to "go *Clockwork Orange*" on a fax machine (as hilariously depicted in Mike Judge's movie *Office Space*). However, we should not let our frustrations blind us to the inconsistencies involved. In other words, we should come to terms with the fact that reinvoking the destructive passion of the Luddites, over and against today's technology, entails smashing the guitar, strangling the song, and slashing the paintings which decorate the commune, as much as destroying televisions and factories.

As Jacques Barzun put it (speaking to an apprehensive audience of pioneering electronic musicians in 1961):

> Most people of artistic tastes share the widespread distrust and dislike of machinery and argue that anything pretending to be art cannot come out of a machine: art is the human product *par excellence*, and electronic music, born of intricate circuits and the oscillations of particles generated by Con Edison, is a contradiction in terms. Here again the answer is simple: the moment man ceased to make music with his voice alone the art became machine-ridden. Orpheus's lyre was a machine, a symphony orchestra is a regular factory for making artificial sounds, and a piano is the most appalling contrivance of levers and wires this side of the steam engine. (2004, 369)

It is therefore necessary to understand that there is no *human* way of getting back to nature.[5] It is too late. In fact, it always *was* too late, and we must accept the fact that the arts and crafts which we celebrate merely comprise the other side of that same coin, struck with images of the microscope and silicon chip. *Il n'y a pas hors de techne*. Becoming a neo-Luddite is a false solution and one which flies (or rather *sits down*, stubbornly, like the three monkeys) in the face of what *is*. The recourse to "harmony" is a nostalgia. (And as Pynchon reminds us, nostalgia is like seasickness: Only the hope of dying from it keeps us alive.)

Writers of many different stripes often quote the final sentence of Donna Haraway's "Cyborg Manifesto," where the author states her preference for being a cyborg over a goddess. But more relevant for the current discussion is her penultimate sentence, which insists that critical thought "means *both building* and destroying machines, identities, categories, relationships, space stories" (1991, 181, my emphasis). A little earlier in the piece, she asserts, "The machine is not an it to be animated, worshipped, and dominated. The machine is us, our processes, an aspect of our embodiment. We can be responsible for machines; they do not dominate or threaten us. We are responsible for boundaries; we are they" (180). And, naturally, we cannot forget this famous line: "The cyborg is our ontology; it gives us our politics" (150).

No doubt it takes a great effort to unlearn deeply embedded distinctions such as the one which places rationality in opposition to play: science on one side and art on the other. Yet, our machines, after all, are just as likely to be corkscrews, swings, sex toys, and turntables as iron lungs, elevators, guns, and cameras. As a consequence, we must remain vigilant against what Bergson so accurately depicted as "so many lame answers to badly stated questions" (1988, 45).

But then the issue becomes a question of transcending Luddite tendencies while resisting techno-determinism or fatalism. What to do when we realize the misconceptions supporting simplistic oppositional strategies? Well, the next step is to come to terms with the *techne within* and acknowledge that the presence of technics is as essential to us as the organic pump which keeps us alive and tells us when we are in love. One cannot slice up a continuum, for it *continues* despite our protestations.[6] The "environment" that we seek to save includes Styrofoam and carbon monoxide as much as koala bears and rainforests. And "pollution," if we want to be rigorous with the term, includes our sewage systems, our museums, our democracies.

Of course, it is in no way misguided to seek to dismantle nuclear power stations or control the use of pesticides. However, this should not be done on the grounds that such technologies are out of step with Nature—otherwise, we will have to dismantle the dams constructed by beavers, as well. "But these objects are *found* in nature," comes the protest. "We have no quarrel with the constructions of these cute furry creatures." To this comes the response that *all* objects are manufactured, whether by beavers, volcanoes, or humans. What is it, then, that prompts the neo-Luddites to draw a line before humans and pronounce, "All hereafter is artificial"? Beavers weren't here at the origin (unless one wishes to resort to smuttiness—Gustave Courbet's portrait of the "Origin of the World," as acquired and coyly displayed by Lacan). Who, other than Darwin, gives beavers the right to build dams? Isn't this a glitch in the Gaia? The beginning of a slippery slope to China's gigantic hydroelectric hubris?

No. From both a biological and an ontological perspective, factories are merely the latest incarnation on the Möbius strip of what exists here on earth. This is not to indulge in the passivity of "What will be will be" but to refigure the way in which we think these incarnations and our complicity with(in) them. In Deleuze's famous example, the stirrup is not simply a technological evolution but part of an assemblage which includes the horse, the leg, and the lance. This more holistic view of such developments is necessary—not in order to understand the complex machinery of a functioning of a whole, as in the metaphor of God the Watchmaker, but in order to appreciate the extent to which "the whole" is beyond "our" grasp. There never will be any Grand Unifying Theory of Everything (unless we scale down our definition of "everything" to human size, something attempted on a daily basis in the laboratories of both the sciences and humanities). Closure and completion are not an option, and neither are the various "carrot-on-a-stick" teleologies which have sustained much of the research and development of the past fifty years.

Thus, to insist, in the spirit of universal communion, "We are all human," is to say less than nothing.

As we have noted several times, people are comfortable talking about computers, mobile phones, and x-rays as technology; however, they are far less willing to consider bread, language, and even "love itself" within the same domain.

So, let us take the example of "gaydar," an intangible technology of detection that we all "possess" to differing degrees. Skeptics may maintain that this is purely metaphorical, that we don't "actually" have a radar that picks up the presence of (fellow) homosexuals.[7] However, this is to reduce the essence of technology to hardware or electronics; it is to mistake a subgenre for a genre unto itself. Gaydar is a capacity, a technique, a sensitivity which reads the signals given off, either consciously or unconsciously, by a fellow human being. It is often a fluency in body language, philology, or the semiotics of fashion and thus qualifies (in Heidegger's sense, at least) as a *techne*.

Let us also consider my brain.

When I say "my" brain, I actually mean, my laptop's hard drive. Any of my own thoughts which I deem as semi-interesting—whether it be an idea for a movie, a bad pun, or a title for a conference paper I'll never give—I log into an appropriate file within my *My Documents* folder. Describing my computer as my brain barely qualifies as metaphorical: "My thoughts" are stored in the magnetized digits of my ThinkPad far more effectively than in my organic cortex. Thanks to my personal computer, I have, in a sense, total recall of those nebulous musings that flash across my consciousness. If I lose track of where I put a prize piece of information among my hundreds of different files, I can use a search facility to locate it.

It is a commonplace to note that we have become so reliant on this kind of technology that it becomes a source of wonder as to how the previous generation remembered anything, or how they managed to write manuscripts without a cut-and-paste feature. It is also a commonplace to complain that we are "too reliant on technology in general," let alone emerging information technologies. My response to this complaint is to ask the following question: If a man has a stroke, do we say he was too dependent on his brain? Similarly, we are quick to say that we are being "bombarded" by media and technology. But do we say that fish are being *bombarded* by water?

It is this insistence on limiting the definition of *technology* to hardware that blinds us to the implications of the so-called technological revolution (or to what Heidegger would call its "essence"). When I say that my portable computer is my brain, I see this as a simple cyborgian isomorphism, like that of inline skates or hearing aids. It is fundamentally instrumentalized and completely "natural." If my hard drive happens to "die" and I have not backed up my documents recently, then I will have lost a great deal of "my own" memories. It would not be *humanly* possible to recall all my notations in any given week, but I could access them easily with a screen or printer. I am a cyborg. (And if you are reading this, then you are one, too.)

Perhaps we should consider yet another example, in the likely scenario that the reader remains unconvinced. A fisherman in southern China uses cormorants as a tool of his trade. Naturally enough, these birds catch the fish. Before a cormorant can actually swallow a fish, though, the fisherman catches the bird from his raft and squeezes it until the fish plops out onto his boat. This seems as cruel as it is allegorical, and although no "hard" technology is involved (except the raft), it is a highly technological arrangement. Moreover, it is this kind of arrangement which has inspired more modern-minded people to design "living machines" which use plants, fish, and microorganisms to break down toxic wastes from despoiled rivers and wetlands.[8]

When we broaden the definition of the circuitry involved, the machinic pulse of the world becomes more apparent everywhere we turn. For instance, it is revealing to question the motives behind the attempt to eradicate drugs in sport. What assumptions underlie this media hysteria, as well as the moralistic detection process which accompanies and encourages it? The backlash against steroids and other "performance-enhancing substances" relies on the quaint notion of a human subject somehow unsupported by technology, as if training on gym equipment, wearing Kevlar bodysuits, or eating high-protein meals were somehow more "organic" or natural than so-called drugs.

And what makes this stance even more ironic is that today's superdrugs are usually synthetic hormones or high concentrations of what the body *already produces* when left to its own "devices," blurring the lines between the natural and the artificial even further. So, perhaps it is time to drop the reactionary attempt to separate the athletic human subject from the machine—as if we were still in the days of the 1936 Olympics in Hitler's Berlin—and embrace the ambivalence of our dependency on certain tools: internal, external, physical, chemical, emotional, and psychic.[9] (Then there would be no ideological irony underlying the Paralympics, which allows "disabled" athletes to compete against each other with technological prostheses *on the condition* that they preserve the spectacle of the human triumphing against both adversity *and* the NASA-designed crutches that carry them across the finish line.)

Spiderwebs, bird nests, beaver dams, chimpanzee tools are all portents of those instruments fabricated by humans. And so it appears that humans necessitate the *accelerated* appearance of technology, recalling McLuhan's strategic observation that humans are merely the sexual reproductive organs of the machine world, or Evelyn Waugh's belief that "Man is . . . never happy except when he becomes the channel for the distribution of mechanical forces" (1943, 159).[10]

The situation, however, is more nuanced than this. Furthermore, it is not a matter of simply reversing the hierarchy but of seeing the "mutual illumination" between humans and technology. For, as with Deleuze's essence, there is no technology outside those entities which evoke, manifest, and incarnate them. Just as essence individualizes and can appear only in such forms, technology requires humans to more fully come-to-presence, and vice versa. This is why the word *computer* was first applied to people, before calculating machines came along and usurped them. To pose the question concerning technology in terms of precedence (of chickens vs. eggs) is to confuse product for process in the interests of reifying power. That is to say, it matters little whether you believe "humans are still in control" or that "machines are rampaging out of control," as long as you see these constitutive terms as autonomous and/or antagonistic.

The above point is as tempting to dismiss as it is crucial, so it may be worth lingering just a little longer on the dynamic relationship between technology and that quaint philosophical category known as "Man."

"The essence of technology," writes Heidegger, "is by no means anything technological" (1977a, 4, 20). From this assertion many unfamiliar and promising insights emerge. The human tendency to view technology as either neutral, or instrumental, or diabolical is—according to Heidegger—to mistake the woods for the trees (or more precisely, to mistake a particular tree for "that which pervades every tree, as tree"). For Heidegger, then: "There is no demonry of technology, but rather there is the mystery of its essence" (28).[11]

The "anthropological" definition of technology therefore blinds us to the essence of technology—that being its catalytic role in revealing the truth of Being to those who have lost any inkling of what this abstract term might mean. As modern humans, we are so distracted by "the unreasonable demand" of tapping, transforming, controlling, and storing nature in standing reserve that we are largely deaf to the call that lies at the source of all this frantic activity. Were we to pay heed to the wider logic within technology—what Heidegger calls Enframing [*Gestell*]—then we would be alerted to the "saving power," the silver lining, of "the monstrousness that reigns here" (16).

"But suppose now that technology were no mere means," posits Heidegger. "How would it stand with the will to master it?" (5). Again, such a premise completely changes our relationship to the objects that surround us, or at least the way we consider this relationship.[12] Rather than seeing ourselves as the center of the magic machinic circle, we should see ourselves in an almost cosmic dialogue with the forces of appearance; of that "unconcealment" which the universe brings forth [*techne*].[13]

To repeat: It matters little whether the agent of this appearance is Mother Nature or Bob the Blacksmith, since both are involved in the same essential labor of bringing something forth into the world, albeit in different registers. "Does this revealing happen somewhere beyond all human doing?" asks Heidegger. "No. But neither does it happen exclusively *in* man, or decisively *through* man" (24).

He continues:

> Man can indeed conceive, fashion, and carry through this or that in one way or another. But man does not have control over unconcealment itself, in which at any given time the real shows itself or withdraws. . . . Yet precisely because man is challenged more originally than are the energies of nature, i.e., into the process of ordering, *he never is transformed into mere standing-reserve*. Since man drives technology forward, he takes part in ordering as a way of revealing. But the unconcealment itself, within which ordering unfolds, is never a human handiwork, and more than is the realm through which man is already passing every time he as a subject relates to an object. (18, my emphasis)[14]

The popular technophobia of *The Matrix* trilogy (1999, 2003), whereby humans are reduced to bio-batteries which power the giant machine, thus mistakes the core of the isomorphic issue.[15] Although Heidegger would be the last person to discount the presence and power of modern reification and alienation, he would never go so far as to claim that we have been reduced to "merely cogs in the global machine," since we have the potential to question, listen, and respond to what this machine (that we ourselves, it is true, comprise a privileged part of) is trying to tell us.

So, while our Bavarian guide can claim: "The machines and apparatus are no more cases and kinds of Enframing than are the man at the switchboard and the engineer in the drafting room" (29),[16] Heidegger looks forward to the day that technology reaches a saturation point and implodes into its own reversibility, encouraged by our eventual acknowledgment that technology cannot be reduced to "planetary technics." For it is only when we finally cease behaving like Nature's parole officer, and renounce the "delusion" that the world of technology is of our own making, that we can transcend the distracting sideshow "calling for an ethics of the technological world." "Let us at long last," pleads Heidegger, "stop conceiving technology as something purely technical, that is, in terms of man and his machines" (2002, 34).

And why? Because the human world and the technological world are more than simply interdependent; they are inter-*determining*. Machines and humans constitute a Möbius strip—a twisted strip with two sides, yet consisting of only one plane. With this simple and elegant metaphor, we can appreciate the difficulty of finding the exact point of the "twist" which leads to the illusion of two opposite sides. This is the Möbius striptease, which has thrilled the progressives and disgusted the technophobes since the consolidation of the industrial revolution.

At this point, however, the lobbyists and the materialists may cry foul, claiming that all this talk of essence and Möbius strips is all very well, but it does not deal with the *here and now*, that being the injustices and humiliations delivered daily by the cold hands of technology. (What Heidegger calls "the profundity of the world shock that we experience every hour" [1977d, 157].) But just as the lofty intentions of ideology and discourse often create something as visceral and vulgar as torture, "concrete machines" are at the mercy of, and operate according to, the abstract machines which generate, code, and create them. Hence, every *actual* machine is saturated and infused by the logic of the *virtual* machine.

Quite simply, then, we are slaves to the notion that we are masters.

The Algebra of Need

> To speak of love is in itself a jouissance.
>
> —JACQUES LACAN, *On Feminine Sexuality*

Imagine a storybook girl sitting in a field of buttercups—perhaps even Irigaray herself at age 16—sighing and smiling and plucking the petals of a flower. She begins to pull the petals away from the stamen, one by one, in the ancient ritual of "He loves me, he loves me not": a game of romantic roulette, as well as an intimate echo of the Hellenic oracle. But is this flower innocent? Can we consider it as something separate from, or outside of, the machinic assemblage of the lover's discourse? The girl seeks a sign—or better yet, seeks to *produce* a sign—from the great text of Nature, in which she imagines her love has been written. Moreover, in doing so, she transforms the flower into a tool of her desires.

Such an image exposes the ease with which everything a human being touches turns to culture (not so much the Midas touch, but the Hephaestus touch).

Nevertheless, in another twist of the tale, we as hubristic mammals are of course tissues of the natural world and are incorporated within it, no matter how much we try to manipulate, deny, or transcend it. So, distinctions between nature and culture are somehow beside the point when discussing such matters—meaning *all* matters, since it is we who are doing the discussing, and we can discuss nothing but ourselves (that is, "the world"). This statement is not made in order to place humans at the center of a sphere which includes everything else but to note that "Being cannot *be* anything but being-with-one-another, circulating in the *with* and as the *with* of this singularly plural coexistence" (Nancy 2000, 3). And the most enduring name given to this *with* is *love*.

The *Oxford English Dictionary* partially defines "love" as "That disposition or state of feeling with regard to a person which (arising from recognition of attractive qualities, from instincts of natural relationship, or from sympathy) manifests itself in solicitude for the welfare of the object, and usually also in delight in his or her presence and desire for his or her approval; warm affection, attachment. Const. *of, for, to, towards*." Indeed, it is this very movement *toward* that has been the focus of our study: a technological groping toward community.

It is crucial to note, however, that deconstructing the distinction between human and machine is not to reduce those phenomena usually associated with the former (affect, memory, agency, passion, and more) to an effect of positivist technics. In other words, I have not been arguing that love is simply a code, as John Brockman does, when he states:

> "I am in love." The neural impulse does not necessarily bear relationship to the sensory stimulus. Stereotyped neural programs can be activated in any number

> of different situations. "I am in love." Faces, bodies change but the same love remains, the same feeling. Such stereotyped programs are established by prior experience which both encodes and rigidifies the operant activities of the brain, delimiting the range of potential responses. "I am in love." All pleasures, all love exist in the brain. Neural programs. Not heart. (1973, 76)

Such a positivist perspective can only encourage such instrumentalist guidebooks as *Love Mechanics: Power Tools to Build Successful Relationships with Women* (2001) and *The Marriage Mechanics: A Tune-Up for the Highway of Love* (2002)—only two of countless such titles. Indeed, we don't have to be as sophisticated as Heidegger to realize that such an attitude is "merely technological"—insofar as it is caught in the truthless Being of calculation. (Man-as-*metron*: the one who not only measures according to his own scale but displaces such unsavory habits onto the rest of the world.)[17]

To welcome the role and legacy of *techne* is not to automatically embrace an instrumentalist technics based on principles of ratiocination, since this ancient concept sees no intrinsic difference between the manufacture of art and industry, poetry and science. Plato's Agathon tells us that love *is techne*, since the arts of medicine, archery, and divination were discovered by Apollo under the *guidance* of love and desire. Melody, musical instruments, metallurgy, weaving, and even empire are all due to love, the *inventor*. It thus begins to dawn on humans that they are the mediums through which love invents and creates itself.[18] As Stiegler says, we must recognize a logic driving technology itself.

For her part, Kaja Silverman maintains—convincingly, I might add—that when we look at our surroundings, "in the most profound and creative sense of that word, we are always responding to a prior solicitation from other creatures and things." Moreover, this visual response is intrinsically libidinal: "What the world of phenomenal forms solicits from us is our desire" (2000, 144). In other words, things appear (both sentient and less so) in order to "invite us to make them part of our singular language of desire—to make them components of the rhetoric through which we 'care' " (144). Stripped of the sophisticated logic behind her belief that perception *is* passion, Silverman is led to state in no uncertain terms, "The world does not simply give itself to be seen; it gives itself to be loved" (133).

But what happens when the perceiving subject is not human but a bona fide cyborg, like Steven Spielberg's and Stanley Kubrick's bastard love child–

robot David? Providing we have the patience to wade through the cloying quicksand of *Artificial Intelligence: AI* (2001), it is possible to see in this film the tensions of an epoch which is beginning to acknowledge that love is a cultural code, in Luhmann's sense, and yet willing to invest millions of dollars and excessive emotional energy in disavowing this same realization (or at least in insisting that there is an invisible remainder to the lover's discourse).

Without delving too far into the narrative itself, it is sufficient to note the uncomfortable affinity between the behavior of its protagonist, the robot-boy David, with John Brockman's preprogrammed perspective on love quoted above. For as the film theorist Drehli Robnik maintains: "*A.I.* is so obviously not about questions of robotics and neurosciences, a more adequate title might have been *Artificial Love*. The problematic gift which robot-boy David is blessed with, after all, is not his outstanding intelligence, but his capability to love. The surrogate-child's pre-programmed love for his human adoptive mother sets the tone for the film's affective register; love, in all its mind-blocking, sentimental naiveté, is stubbornly acted through—like a program" (2002). And yet, as with the film's less syrupy predecessor *Blade Runner* (1982), the love that the replicants seek is something which escapes the logic of the simulacra. Indeed, it is as if Walter Benjamin's famous comments concerning the aura have been enlisted by the cinematic apparatus and applied to people, or rather to androids, instead of to works of art.

On the one hand, the age of mechanical-digital reproduction has robbed the once authentic, originary object of its unique status. On the other hand, this very same "decay" of the aura anticipates and allows the radiating half-life of exchange value and commodity fetishism. Thus, the halo is still there; however, it can be seen only under the cultural equivalent of ultraviolet light (i.e., consumption, the logic of the Spectacle, and so on). And if we replace the word "object" with "subject" here, the stakes become even clearer. Thus, when Heidegger insists that "there where the danger lies, grows the saving power also," Hollywood responds by trundling out their most popular salvage expert, Love. *Technology may be robbing us of our essential humanity*, says Spielberg, *but the power of love will be our redemption*. (This is a sentiment shared by *The Matrix* [1999], specifically in the reverse "Sleeping Beauty" moment, wherein Trinity saves Neo from brain-fry by invoking and executing the life-affirming command "I love you.")[19]

Steven Shaviro discusses "the algebra of need" (William Burroughs) in his book *Connected* (2003), and elsewhere, in order to wrench desire from the withered, yet surprisingly powerful, grasp of this kind of Romanticism. For instance, he sees an emerging posthuman incarnation[20] of erotic pleasure in Chris Cunningham's video for Björk's "All Is Full of Love," suggesting that "when we have become posthuman cyborgs, we will still have some sort of bodies. We will still have tenderness and yearning, and still need to make love" (2002, 30).

Moreover:

> Usually, we think of machines as being uniform in their motions. They are supposed to be more rigid than living beings, less open to change. But "All Is Full of Love" reverses this mythology. It suggests that robots might well be more sensitive than we are. They might have more exquisite perceptions than we do. They might respond, more delicately than we do, to subtler gradations of change. (103)

This kind of *jouissance* is a step beyond the predigital dispensation which inspired the 1929 musical hit song "I Wish I Had a Talking Picture of You." It also transcends the kind of media archaeology which would consider, say, *The Kama Sutra* as an analog and indexical database for erotic technologies of togetherness.

"Mediated desire," as explained by René Girard, misleadingly suggests the possibility of an *un*-mediated desire: an impossible situation, given that desire itself *is* pure mediation. (I desire what the other desires, which itself is a reflection of someone else's desire, *ad infinitum*.) Thus—in terms no doubt familiar to those who have followed the full logical tangent of this book so far—this mediation is not a process that *connects* the two "points" doing the desiring (that being the lover and beloved), but a movement that simultaneously *creates* the two amorous "switching points" being mediated. In this sense, we are indeed desiring machines—but machines created by the network of which we are always already a part.

As emphasized at the beginning, the enigmatic performative speech-act known as "I love you" has seduced most minds into an inquiry into the nature of the verb in the middle, rather than the personal pronouns which sit on either side of it, separated like awkward teenagers by a family-approved chaperone. Who, in other words, are this "I" and this "you" that claim to love? But we need not resort to existential metaphysics to answer this question, since we ourselves often ask this about our partners in a very pragmatic, mundane kind of sense (as in "Who the hell did I marry?" or "You'll never understand *me*").

Seeking to capitalize on this structural unknowability of one's beloved are a slew of software solutions to the clandestine online behavior of one's partner. (And as we well know, yet another Möbius strip is the misleading distinction between online and offline, since it takes only one email to arrange a lunch date at an off-ramp motel.) No matter how inspired by the everyday agonies of jealousy and insecurity, these invisible surveillance programs—boasting names such as Keylogger, I Am Big Brother, Spybuddy, and (we can only hope) the Spouse Trap—nevertheless do have significant ontological effects. One's partner is no longer the immediate flesh-and-blood creature that one seeks to "hold onto" but the nocturnal source of unfathomable, intangible, keyboard-based professions of love to someone known only as BayArea_Betty. Someone's "personality" is—to use Heidegger's formula—unlocked, transformed, stored, distributed, and switched from home and hearth to the cold global mediascape which lurks on the other side of a modem. And in doing so, the subliminal surveillance that has always been a part of the domestic sphere plays out on a minor scale the diffusive logic of what Mark Poster calls "the superpanopticon":

> Foucault argued that the subject constituted by the panopticon was the modern, 'interiorized' individual, the one who was conscious of his own self-determination. The process of subject constitution was one of 'subjectification', of producing individuals with a (false) sense of their own interiority. With the superpanopticon, on the contrary, subject constitution takes an opposing course of 'objectification', of producing individuals with dispersed identities, identities the individuals might not even be aware of. The scandal, perhaps, of the superpanopticon is its flagrant violation of the great principle of the modern individual, of its centred, 'subjectified' interiority. (1996, 291)

Hence the crucial point: "A politics that circumscribes freedom around the skin of the individual, labeling everything inside private and untouchable, badly misconceives the present-day situation of digitized, electronic communications" (291). In other words, whether we are confronted by the FBI for our terrorist activities or by our spouses for our infidelities, it is historically naïve to claim that "our" rights as an "individual" have been "violated" (with all the assumptions of autonomy and coherence that these concepts entail).

The paralegal mechanisms put in place to facilitate global capital have forced a mutation in that concept once dubbed "justice," forcing us to reassess the relationship between *individual*, *sovereignty*, *human rights*, *(state) violence*, and *freedom*. And while it would take another book to detail the intimate link between contractual obligation in both love and commerce, it is increasingly clear that what Poster calls "interpellation by database" has dramatically reconfigured the status of those beings represented by a signature. Put simply, snoop software and online databases are "restructuring the nature of the individual," functioning as "performative machines, engines for producing retrievable identities" (287–88). And as we have just noted, in most cases "the individual is constituted in absentia, only indirect evidence such as junk mail testifying to the event" (288).

Under such conditions it makes little practical sense to espouse knee-jerk defenses of the integrity and privacy of individual identity. For as David Lyon bluntly puts it: "Recognizing people as unique identities . . . makes their control easier" (1994, 79). When we inflate this understanding to the level of society as a (conceived and illusory) whole—which we must, for both political and ontological reasons—then it becomes prudent to agree with Poster's claim that "in the era of cyborgs, cyberspace, and virtual realities," all appeals for community "must take into account the forms of identity and communication in the mode of information, and resist nostalgia for the face-to-face intimacy of the ancient Greek agora" (1996, 291).

How exactly we "take this into account" is the subject of many current books, articles, interviews, and conferences. However, the majority of these fall into the oversimplified binary of a centered liberal-humanist subject versus a decentered postmodernist and posthuman(ist) subject. As Steven Shaviro so perceptively warns:

> Psychoanalysis is most often taken as a deconstruction of the supposedly unitary bourgeois subject, and as a liberation of the forces repressed within it. I want to suggest that this is far too limited a view; the decentered psychoanalytic subject is not something that comes after the Cartesian, bourgeois subject, but something that is strictly correlative with it. In contrast, a new, posthuman subject will have to point away from Freudian and Lacanian conceptions of decentered subjects, as much as from the unitary Cartesian one. The whole frame of reference will have to be different. We will have to understand the body/mind in other terms, according to the play of other structurations and other forces. The current computer-based analogies to the mind, common among cognitive scien-

> tists, are as desperately simplistic as the old Cartesianism was; but it will need to be answered and complexified by something that responds to new digital models as intimately as psychoanalysis responded to the Cartesian notion of a unified ego. (2002, 29)

The virtue and value of Agamben's *whateverbeing*, I would argue, is that it both avoids this stagnant binary of centered versus decentered and opens new pathways to theorizing what could be called the postsubject subject. When combined with Deleuze's strategic deployment of the term "essence," as well as Nancy's refinement of "being singular plural," we have a formidably sophisticated and nuanced theoretical machine for producing poetic revelations and political responses concerning the immanent interdependency of love, technology, and community.

EIGHT

Asymptotic Encounters: Love Freed from Itself

> Whenever I happen to be in a city of any size, I marvel that riots do not break out every day: massacres, unspeakable carnage, a doomsday chaos. How can so many human beings coexist in a space so confined without destroying each other, without hating each other to death?
>
> —E. M. CIORAN, *History and Utopia*

> It's so hard to go into the city,
> because you wanna say 'Hello' to everybody.
>
> —CAT POWER, "Colors and the Kids"

> If this world were not endlessly crisscrossed by the convulsive movements of beings in search of each other . . . it would appear like an object of derision offered to those it gives birth to.
>
> —GEORGE BATAILLE IN MAURICE BLANCHOT, *The Unavowable Community*

> —But where do these nobodys live, and how?
> They live in the city, and they live in "*everydayness.*"
>
> —KAJA SILVERMAN, *World Spectators*

In that particular genealogy linking Baudelaire to Michel de Certeau via Benjamin, Musil, and *The Man with a Movie Camera* (1929), the city is figured through a kind of semichoreographed *ballet mécanique*. While we may not be able to talk of harmony, there is certainly some kind of order in the to-ings and fro-ings of the city, especially when viewed from the 110th floor of the World Trade Center.[1] Various patterns emerge, which may or may not be predictable but whose flux can be traced through the dynamic assemblages that comprise the condition of their movement. Specific strange attractors come into play, simultaneously within, around, and between the places and spaces of the urban system.[2]

To map this hypermobility, we can artificially subdivide these movements into two categories: the "macro-movements" of gentrification, de- and reindustrialization, highways, private and public transport vectors, curfews, and official calendars; and the "micro-movements" of impromptu parades, buskings, burglaries, seductions, meetings, and meanderings.[3] Such a distinction is "artificial" because, when pressed, there are no adequate criteria to judge the macro from the micro, since busking can be the direct result of deindustrialization, motivated by the rent inflation of gentrification and supported by the spaces of public transportation. As with all assemblages, these heterogeneous elements cut across each other, hold, twist, turn, and unravel, according to the currents and conditions of their own emergence. As Deleuze often reminds us, reterritorialization follows swiftly on the heels of deterritorialization—and so much so that it is too simplistic to view it as a chronological procedure. Rather, it is an immanent process, rarely grasped in flux but glimpsed through an uneasy marriage between systems theory and schizoanalysis.

The difference between Baudelaire's city and our own is that it makes less sense to begin with the subject, no matter how marginalized or alienated he or she may appear to themselves or others. Such a figure no longer has the critical distance or historical perspective to appreciate urban ironies, juxtapositions, and shocks. They simply *are* urban ironies, juxtapositions, and shocks—human pollen landing on concrete. Yet, as I have tried to demonstrate in the previous chapters, collectively, the *populi* are the site of a certain movement or tendency toward the coming community (a "multitude" in the sense offered by Alliez, Hardt, Negri, and others in their journal of the same name).

Our foremost poet of the postmodern *flâneur* is Wong Kar-wai.[4] Those of his films set in the contemporary moment—most notably *Chungking Express* (1994) and *Fallen Angels* (1995)—are sublimely stylized homages to the asymptotic nature of modern love, in which the "protagonists" barely qualify for the term and drift past each other as if to explicitly enact the dictum "There is no sexual relationship," or indeed, no relationship, period.

Cultural critic Ackbar Abbas notes that in Wong's universe, "all appointments are dis-appointments" (50). As a consequence, these films (set in Hong Kong) depict a particularly reified version of that "eroticism of disappointment" which also preoccupied Musil, Broch, and Proust. The task of the viewer, however, is to avoid sentimentalizing these random, asymptotic encounters along the lines of "ships that pass in the night"—even if Wong himself occasionally falls into this trap (as happens with *In the Mood for Love*).[5] For if we manage to steer clear of this temptation, it then becomes possible to accept Wong's characters as postmelancholy heralds of whateverbeing and thus to see the lover's discourse with a fresh pair of eyes. "Seeing something simply in its being-thus—," writes Agamben, "irreparable, but not for that reason necessary; thus, but not for that reason contingent—is love" (1993a, 106).

Speaking of the ongoing "codification of intimacy," Niklas Luhmann writes:

> The idea of a long, drawn-out pining for fulfillment seems ridiculous. And yet, the ties and impressions left on a person by a sexual relationship lead only to unhappiness. The tragedy is no longer that the lovers fail to find each other, but rather in the fact that sexual relationships produce love and that one can neither live in keeping with it nor free oneself of it. (1998, 160)

Fallen Angels, to take the most typical instance in Wong's catalogue, does not disavow the fact that ties form between people, sexually or otherwise. However, these are always fleeting, contingent, fragile, and yet somehow stubborn.[6] People do not so much yearn for each other as exhibit a chthonic kind of yearning, arising from the city street itself, which gives them the momentum to find provisional and impromptu opportunities for a kind of community.[7] (For instance, there's the brief solidarity that forms between a crazy man and the family whom he hijacked with an ice cream truck, or between the same crazy man and the pig carcass he decides to massage simply for effect.) And so, when *Time Out New York* says of Wong Kar-wai's films, "Nothing really happens, but everything seems at stake," we can nod our heads in the understanding that this description could indeed be applied to life in general.

The temptation then also arises to consider the pseudo relationships depicted in *Fallen Angels* and *Chungking Express* as embodiments of Bauman's "liquid love," in which no given person has the "stickiness" (to borrow an internet marketing term) to attract a sexual partner over a prolonged amount of time. In an era populated by men and women without qualities (or at least scrolling, chameleonesque, inessential qualities), we must strive for anonymous "hits" rather than a loyal fanbase.[8]

Take, for instance, this scene from *Fallen Angels*, where a young woman, "Baby," explains her arresting new choice of hair color:

> *Baby:* Guess why I became a blonde.
> *Ming:* I don't know.

Baby: So no one will forget me.
Ming: Looks great.
Baby: You're making fun of me.
Ming: No, you're really special.
Baby: I've heard it all before.
Ming: From whom?
Baby: You! We were together for some time. I had long hair. You called me "Baby."
Ming: Really?[9]

A general amnesia pervades not only Abbas's Hong Kong but the people who constitute it: "Such is the postmodern, or posthuman, equivalent of love. It's a love in which the subject and the object never meet; even connected by a glance or a stare, they remain apart. Such love is a cool, aesthetic contemplation; it is even 'indifferent to the existence of the object' being loved, which is Kant's definition of aesthetic disinterest" (Shaviro 2003, 73).

And yet, things are not quite as bleak as they appear, since they produce their own halo effect to counter the implicit claim that love is merely narcissism *à deux*. In other words, we encounter a remainder, a not-quite-accursed share, a virtuality which points to an exit of sorts from this kind of grinding, urban, spectacle-saturated alienation. Moreover, this gift for rendering melancholy and longing without the obvious props of romantic sentimentalism is by no means reserved for Wong alone. Heidegger's much-anticipated "inevitable dialogue with the East Asian world" (1977d, 158), comes to us not only in the guise of art house films but also in the novels of Haruki Murakami. The sudden popularity of these texts in the so-called West is, I would argue, symptomatic of our epoch's sensitivity to whateverbeing, especially among a generation exhausted with I-dentity and seeking something closer to a we-dentity.

Murakami is notable for taking the archetypal hard-boiled mouthpieces of Hammett, Fitzgerald, Hemingway, and Carver and shedding all traces of idiosyncrasy or personality from them for his own narrator-protagonists.[10] As such, Murakami's characters truly are men-without-qualities, as they themselves are often the first to admit: "It shocked everyone that Nagasawa chose me, a person with no distinctive qualities" (*Norwegian Wood*, 2000).

Thus, in Murakami's vision of the world, as much as in Wong's, the banal, the everyday, the quotidian provides the spaces in which minor epiphanies (in Deleuze's sense of minor literature) can prevail. Suddenly, an ashtray full of beer can pull-tabs metamorphose into mermaid scales, a family kidnapping becomes an impromptu outing, and a nocturnal joyride on the back of a motorcycle can create a libidinal event in the landscape of the postmodern subject. And while these moments may be exactly that—*moments*—without any longevity to encourage political mobilization against the mechanisms which prevent them from happening more often, they nevertheless allow us to appreciate Bergson's point that moments *continue* long after they've had their "moment."

Interestingly, it is these "Westernized" Asian texts (or rather the dialogue between and among these different cultural hemispheres) that have offered a more nuanced presentation of "inessential commonality."[11] Trinh T. Min-ha, in terms far less compromised than Irigaray, emphasizes the fact that the "nature of *I*, *i*, *you*, *s/he*, *We*, *we*, *they*, and *wo/man* constantly overlap, they all display a necessary ambivalence, for the line dividing *I* and *Not-I*, *us* and *them*, or *him* and *her* is not (cannot) always (be) as clear as we would like it to be" (1986, 56). Indeed, she goes so far as to claim: "You and I are close, we intertwine; you may stand on the other side of the hill once in a while, but you may also be me, while remaining what you are and what I am not" (90). (One need only be a parent, I imagine, to appreciate this kind of distributed identity.)

However, in the spirit of deconstructing simplistic *either/or*s, it is useful to remind ourselves of Barthes's point that while the lover's discourse can seemingly connect us to otherness, it can also painfully condemn us to our psychically isolated singularity. "*I am not someone else:* that is what I realize with horror" (1990, 121). The slippage or disjunction occurs between the ontological and the psychological—that is to say, the way we "are" in agonistic relation to the way we *think* we are (consciously or unconsciously).

For Murakami, at least, our psychological baggage need not sentence us to a lifetime of distracted solipsism:

> Each piece of luggage I open reveals a similar tawdry inventory. Clothes and some few sundry items, all seem to have been packed for a sudden journey. Yet each wants for identifying detail, each impresses as somehow unremarkable, lacking in particularity. The clothes are neither quality tailored items, nor tattered hand-me-downs. They show differences in period, season, and gender, varying in their cut according to age, yet nothing is especially striking. They even smell alike. It is as if someone has painstakingly removed any indication of individuality. Only person-less dregs remain. (1993, 228)

These are the "be-longings" that belong to no one and everyone, attesting to this affirmation by Nancy: "To exist is a matter of going into exile" (2000, 78).

But didn't I just claim that there was some kind of redemption, or at least a potent potentiality, lying dormant within the texts themselves? Indeed I did, and as always, I can be secure in the knowledge that I am not alone in this, for Blanchot also identifies a certain enabling failure in asymptotic encounters—that is (speaking of two asymptotic beings),

> the failure that constitutes the truth of what would be their perfect union, the *lie* of that union which always takes place by not taking place. Do they, in spite of all that, form some kind of *community?* It is rather *because* of that that they form a community. . . . [E]very form of empty intimacy, preserves them from playing the comedy of a "fusional or communal" understanding. (1988, 49)

Indeed, we would be hard-pressed to find a more accurate description of the beauty which animates Wong's *Fallen Angels* or Murakami's *Wildsheep Chase* (2002). Despite the predominant tone of loneliness and boredom in both, there is always the chance of an encounter. While this encounter may not produce a union, a connection, or even a lasting impression, it does give lie to the debilitating sense that we are discontinuous and isolated individuals.

"That is why," writes Nancy, "one would want to separate oneself from love, free oneself from it. Instead of this law of the completion of being, one would want to deal only with a moment of contact between beings, a light, cutting, and delicious moment of contact, at once eternal and fleeting. In its philosophical assignation, love seems to skirt this touch of the heart that would not complete anything, that would go nowhere, graceful and casual, the joy of the soul and the pleasure of the skin, simple luminous flashes of love freed from itself" (1991, 92).

For Nancy, the "problem of the city," then, is directly related to this fact:"Love is thus not here, and it is not elsewhere. One can neither attain it nor free oneself from it" (93). But this is not a drama to be staged simply according to the directions of the individual, since "a consciousness is never *mine*, but on the contrary, I only have it in and through the community." That is to say, "Consciousness *of* self turns out to be outside the self of consciousness" (19).

Brutally Intimate: The Ambience of the Street

> When I sit at my ease and look out over some vista, landscape or cityscape, where a play of space meets a play of surface, then imperceptibly I melt, with the myriad associations, into a deeper vista, knowing "myself more truly and more strange." When there are other people included in the scene it works differently, I am no longer spread equally over the field of transparency and opacity. The face and the body of each person that I include gathers and bunches me, each of them holds an infinity into which I could fall if they were open—or even if they are not, through the cracks in their presentation, through what I could loosely call their presentiment. As if to ask, "Where is the first place where 'we' can be?" and then to drop, as in a rapidly descending elevator, right out of time and space.
>
> —DAVID ODELL, *A Rushed Quality*

Such are the quilting points of the street: the semipublic places where we gather, exchange, and disperse not only goods and speech but the very sense of who we are among and in relation to others. Us versus Them. Me versus You. Ultimately, it comes down to this: a question of belonging.

David Odell paints the process in the following terms:

> This city is like a vast coral, its myriad cells and chambers formed by the psychic projections of the even more numerous souls that have at any time taken body within it. In spite of all the cultural differences this collective and intrinsically mythic life of the city imprints a sort of family resemblance on all its physiognomies without lessening in any way the extreme quality of each individual expression. "What have they done with their lives?" The title of a book of which each page is a face.[12] (2001)

In the introduction, I mentioned the phantasmagoric quality of Juergen Teller's photography book *Go-Sees*, which features the parade of hopeful models who have knocked at his door. Teller followed this collection with a project entitled *Tracht* (2001), which is described on the French branch of Amazon.com in the following terms:

> The only photographer ever given exclusive access to all the contestants of a Miss World event, Teller spent eight hours in one evening taking brutally intimate head shots of all 88 contestants. Shot away from the stage and removed from the spotlights, the contestants—with caked on makeup, unwavering smiles, and bright eyes—start to appear interchangeable, like a carnival of culturally diverse Barbie dolls requiring lapel pins to indicate country of origin. *Tracht* is

> German for "costume," as in folkloric dress. *Tracht* is Teller's take on the dressing, the commodification beauty takes when pushed, much like the rest of fashion, to the extreme. The end effect of the ruthlessly efficient globalization of image manufacturing makes appearance artificial and style generic; what is left, then, but to extract its mesmerizing effect in the form of a sumptuously attractive artist's book, where one can truly revel in the mysteries of a world where Miss Israel looks more Californian than Miss USA, who in turn looks more Middle European than Miss Estonia, who looks more Gaelic than Miss Ireland, who looks more Moorish than. . . . Yet throughout, one cannot avoid noticing a certain . . . sameness to these visages. *Tracht* deftly captures the costume of institutional modern beauty, and it is a thing to behold.[13]

No wonder, then, that the respected journal *Lacanian Ink* used Teller's pictures (specifically, ones of Miss Sweden and Miss Guatemala) for the cover of number 18, dedicated to the phenomenon of sexual *non-rapport.* In the editorial for this issue, Josefina Ayerza states: "Whether you like her or not Miss Universe will hold the scepter which represents the desire of all men. Yet after a year a new flower will bloom, rekindling the desire of all men, for a whole new year. And this is the fate of the contemporary Aphrodite. Will she ever be the One?" (2001, 3).

Beneath the publicity hype, there is no denying that Teller seems particularly sensitive to the carousel of whateverbeing which revolves around his unflinching camera. Moreover, Teller enthusiastically contributes to the proliferating excess of this phenomena, as if he alone has the power to push the postmodern, postmillennial spectacle beyond all subjectivity, past the point of personality. Indeed, the creepy substitutability of his photographic subjects captures the vacuity behind the daily routine of the urban hipster—this current, curious, and monstrous generation comprised of people who have about as much personal coherence as Jim Carrey's cable guy, yet the laser-sharp narcissistic focus of Snow White's evil stepmother.

Agamben maintains that

> the absurdity of individual existence, inherited from the subbase of nihilism, has become in the meantime so senseless that it has lost all pathos and been transformed, brought out into the open, into an everyday exhibition: Nothing resembles the life of this new humanity more than advertising footage from which every trace of the advertised product has been wiped out. The contradiction of the petty bourgeois, however, is that they still search in the footage for the product they were cheated of, obstinately trying, against all odds, to make their own an identity that has become in reality absolutely improper and insignificant to

> them. Shame and arrogance, conformity and marginality remain thus the poles of all their emotional registers.[14] (1993a, 64)

This particular set of circumscriptions, which for the sake of argument we shall call Romilly syndrome, is most evident in the consumers of sterile lifestyle magazines,[15] who strive to become part of their favorite advertisements, and who live without reservation within and through the spectacle, identifying solely with "types" and tableaux, rather than singularities and idiosyncrasies. Once these human mannequins accomplish and inhabit recognizable "spaces," however ("space" rather than "place" being one of the keywords for this generation and its aspirations), they suddenly flounder. It is as if sufferers of Romilly syndrome become lost once they achieve their ideal scenario, because they do not have the resources to make anything of it (that is, to "bring something forth," in Heidegger's terms). They feel like a stylish model who cannot relax or surrender to the role until they work out the brand or service that they are supposed to be advertising. The admittedly pompous question "What is the meaning of life?" thus becomes twisted to ask, "What is the commodity that underwrites/sponsors my *raison d'être?*" (hence the endless shopping expeditions that nevertheless fail to fill the void of depthless narcissism). "Are we having fun yet?"—this question makes absolutely no ironic sense outside a capitalist context. Hence the frustratingly *forsaken* element which infects all chic gatherings, like a hint of turd mixed in with the pâté.

But this caricature goes beyond a critique of alienation, as there is no assumption of authenticity as an alternative. Agamben and Zizek agree that we can no longer hope for a Heideggerean "turning" toward the essential authenticity of Being, but should rather invest our energies in pushing this logic to the extreme. For it is only in taking "the system" (ideology, capital, enframing, technicity, and so on) literally and at face value that it is possible to sabotage the smooth functioning of what William Burroughs calls the Me-machine. (That same machine functions to neutralize the potential of whateverbeing via the reactionary rhetoric of "individuality," "true self," and "connection," peddled by New Agers, journalists, copywriters, corporate motivational speakers, mobile phone companies, and moral philosophers alike.)

But how to recognize this potential?

Agamben writes, "The task of the portrait is grasping a unicity, but to grasp a whateverness one needs a photographic lens" (1993a, 49). But what about more recent media?

Scott Becker's CD-ROM *Ex-Girlfriends and Other Untouchable Women* is a remarkable document of compulsive cataloguing and classification. This somber, digital parade of ex-lovers is literally a mnemotechnology enabling Becker to work through the trauma of his own personal love life. The parade begins with screen goddesses Gina Lollobrigida and Sophia Loren, then leading through first crushes to the more significant crushing nature of mature heartbreak. (This is probably why the whole exercise, representing months and months of work, is finally dedicated to Becker's dog.) In this CD-ROM we experience an echo of *Go-Sees* and *Tracht*, but in this case the "pearls" are strung together by the arachno-Humbertish threads of personal biography, that being the narrative weaving afforded by hindsight.

Consider a different kind of artwork inspired by heartbreak, namely those melancholy songs or albums which sell in the millions but are dismissed by the critics as being too clichéd. To a great extent, such criticism misses the point of heartbreak in the first place. Heartbreak is experienced as intensely personal, yet its expression can be—in fact usually is—a dialogue with the public sphere, a violent recognition of the necessity of return to the community after a period of self-exile. This is why we tend to express ourselves in clichés when we are heartbroken: because we have been bequeathed the language to describe the situation from those who came before. And this is why we like nothing better, when heartbroken ourselves, than to listen to the music of those who have also been heartbroken.

Thus, "the blues" (in all its multigeneric forms) is a pure passing of the code, which is not to minimize the pain involved but to highlight the obvious fact that no matter how unique one's relationship with the (ex-)beloved, the process and vocabulary of grief has a strict and recognizable form. Nancy is therefore correct in pointing out that the "heart" itself—as a metonym for love—is the phoenix which arises from the ashes, *that we cannot conceive of a heart preceding heartbreak*. Thus, the reflexivity at the "heart" of the system shows itself to be as resilient as it is bruised and buffetted by the lover's discourse, along with those machines of loving grace which comprise its entourage.

It thus becomes necessary to acknowledge a difference between "romanticism" proper and "romanticization" as a mutating outgrowth of the romantic code. This distinction arises from a suspicion that the latter is a valid comportment or strategy—that is to say, a very general symbolic way of negotiating the pleasures and distress of memory, nostalgia, and change (rather than an inherently reactionary worshiping at the altar of human vanity). Clearly connected to aesthetics, *romanticization* may retain the process of self-narration (or "autopoeisis," to Luhmann) without wrinkling in the dirty bathwater of romantic tropes (arcadia/utopia, self-realization at the expense of the other, politically complacent nihilism, and so on). In order to function as a subject, even in an allegedly postmodern world, we must constantly narrate the self in the same way the state must constantly narrate the nation.

This is not to say that the narrative itself changes in tone, focus, form, and content, but that people constantly weave—in fact *are* the very weaving—of future, present, and past: those wistful moments, those sharp, stabbing recollections, those imagined possibilities and memories, as well as all those harsh realities that must be incorporated and processed (what Stiegler calls "infolding"), and then flushed (what the media calls "art," "work," "vandalism," or other modes of sublimation and excretion). Until there is a complete change in the code, Western(ized) subjects will always romanticize themselves and their situation—it is the very medium with which we communicate with the world and with ourselves (see Heidegger's mediation *within* identity or Deleuze's redemptive artistic apprenticeships). But this does not necessarily mean such a process has to be tainted with the more ego-inflammatory and nihilistic residues of the narrative.[16]

The completion of the Human Genome Project, which announced its findings in the millennial Domesday year of 2000, confirmed the suspicions of most cultural critics concerning "race"; that is to say, it is a dubious biological category. The complex maps produced by the project revealed that, at the level of DNA coding, all humans share 99.99 percent of their genetic code with all the others. As one newspaper report puts it, "All that distinguishes an Inuit from a Cockney or an Aborigine, even Britney Spears from Diana Ross, are variations in 300,000 letters in a three billion-letter sequence in the human genome" (Highfield 2001). Here we have a scientific analogue of essence, played out as much in Teller's photographs of Miss World as in the search for a soul mate among the biomass of *Sex in the City*. The "identity crisis" of contemporary global culture can thus be transfigured into an explicit awareness of the coming community, neutralizing Madison Avenue's attempts to cash in on the scattered self-images of modern consumers.[17]

The street has often been associated with the *film noir* aesthetic, the original apotheosis of twentieth-century alienation and of the inverted romanticism of the search for love. According to Frederic Jameson, "separation is projected out onto space itself: no matter how crowded the street in question, the various solitudes never really merge into a collective experience, there is always distance between them" (1983, 131). This search or distance, transfigured and updated by Wong and Murakami, is still figured through the rubric of "connection," a Platonic *telos* as viewed from the other side of the looking glass.

Indeed, as I write, Japanese youngsters have become obsessed with a mobile-phone game called Love by Mail, where the user can flirt with a virtual girlfriend or boyfriend—who in fact is a computer program—"who responds" to SMS text messages. Western commentators are confounded as to why intelligent young people, who are fully aware that they are trying to seduce an algorithm, are so smitten with this libidinal twist on the Turing test.[18] For the systems theorist, however (and to a certain extent, for the psychoanalyst), the joke is on those who find humor in such behavior and who stubbornly labor under the delusion that their own relationships IRL (in real life) are anything *other* than a complex social algorithm. From a Luhmannesque vantage point, the automatic back-and-forth of flirtation is perhaps *the* default setting of modern communication (which explains the disconcerting fact that even my telebanking operator insists on calling me "Dom," before complimenting me on my "beautiful voice").

Traditional conceptions of "the human" as potentially independent from technology become increasingly untenable and obsolete, even at the level of popular culture. Somewhere between youth value and exchange value, between Naomi Klein's logos and Jacques Derrida's *logos*, and between the symbolic order and the *ordure symbolique*, the hum of the spectacle creates its own phantasms of flesh. Whether it be William Gibson's digital pop singer Rei Toei, winking into multiplicitious real-life existence at the end of *All Tomorrow's Parties* (1999), or the numerous attempts to literalize the pixelated Lara Croft into something more palpable, the ontological membrane which separates the virtual from the actual continues to dissolve.[19]

Removing the Thin Diaphragm; or, Finally We Are No One

> A physicist is the atom's way of thinking about atoms.
>
> —ANONYMOUS

How you feel depends on who you're connected to.

—Vodafone slogan

We have not even begun to think "ourselves" as "we."

—JEAN-LUC NANCY, *Being Singular Plural*

As fate would have it, one of our best examples of whateverbeing is explicitly treated neither in Agamben nor in Nancy, but in the 1999 film *The Muppets from Space.* In a truly inspired revisionist Biblical scene, all the animals of the earth attempt to take shelter from the impending flood in Noah's Ark. The idiosyncratic Muppet character Gonzo, however, is barred by the bearded patriarch, who asks, "What *are* you, anyway?"—barely able to conceal his contempt for this uncategorizable creature.[20] Gonzo, thinking on his feet, stammers his reply: "Oh, good question. . . . Technically speaking . . . let's say, put me down as a *whatever.*" Unfortunately, this self-designation fails to convince the zoological gatekeeper, since the criterion for entry (and thus salvation) is clear-cut membership in an established species. So, Noah hands Gonzo an umbrella, just as the raindrops start to fall, and then closes the gangway to any more asylum seekers.[21]

While this scene constitutes a rather playful allegory for the entertainment of children and parents alike, its logic is played out—quite chillingly—in the world of humans and "human trafficking." In a strategy so simple it is sublime, so-called illegal immigrants are beginning to refuse identifying themselves to the authorities, so that Western governments cannot send them "back where they came from." For a brief moment, reported almost indignantly by the BBC in 2003, we have a breathtaking example of Agamben's belief that "[w]hat the State cannot tolerate in any way . . . is that the singularities form a community without affirming an identity, that humans co-belong without any representable condition of belonging" (1993a, 86; 2000, 87). Of course, the politico-juridical machine will soon decide that it can send these whatever-citizens anywhere it likes, anywhere other than "here," as the Australian authorities do on a daily basis with such heartless efficiency. So, while relatively privileged diasporic peoples and expatriates crave some kind of symbolic or surrogate "home," those far more neglected victims of globally regulated capital are prepared to risk anything and everything to *get away* from home—and then, when caught, to *avoid being sent back* home.

Ironically, given the integral complicity of scientific enframing and agenda setting in this sorry state of world affairs, governments turn to political *scientists* for answers. These are the same political scientists who stubbornly push down the path of rationalist technics as the direction to solving the constellation of concerns associated with globalization (i.e., this late age of the world picture). The fact that the problems *themselves* are by-products of technicist policies is never acknowledged, since that would be to admit that Enlightenment Man has been drained of his powers (a prospect more terrifying than the death of God in a high-humanist age like this one).

According to the annals of history, the natural philosopher Archimedes dreamed of a perspective unsullied by subjectivity and other distortions of thought that necessarily accompany the "interested" nature of the individual. Logically enough, this hypothetical point of pure objectivity is now called an Archimedean point. Scientists, political or otherwise, like to believe that they can largely inhabit such a point; or at least use it as a polestar for their disinterested calculations. However, this point will always be hypothetical, and no subject can ever reach a state of pure objectivity, at least not without turning to stone. As this book has repeatedly affirmed, there simply is no self outside of, and apart from, other selves, other forces, other factors, of which "we" are a singular instance. Despite C. B. Macpherson's claim that "human essence is freedom from the wills of others" (in Hayles 1999, 3), one need only recall the tangible force of peer pressure to begin thinking to the contrary. (That is, even if you resist peer pressure, this resistance is inspired by encouragement or discouragement from a different source. For instance, you may refuse a cigarette not because you are "being true to yourself as a nonsmoker" but because you have internalized the advice of your doctor, priest, or grandmother.)

In his book *Seeing Double: Shared Identities in Physics, Philosophy, and Literature*, Peter Pesic explores the suggestive paradoxes posed by quantum theory. Beginning with the established premise that electrons and other elementary particles exhibit no individuality, Pesic then unpacks the irony that we humans (who of course are composed of such elementary particles) refuse to even entertain the notion that we might share some postclassical physical properties with our own building blocks. "[I]t is strangely beautiful," he writes, "that human individuality rests on anonymous quanta" (2002, 148).

In a subatomic deflection of Agamben's paradigm, Pesic notes that every electron is so devoid of distinguishing features that it is impossible to "tag" any particular singularity in order to trace its history. This is what Pesic calls "identicality," meaning that "the members of a species have identity only as instances of that species, without any features that distinguish one individual from another." "Identicality" thus "signifies the exact negation of individuality" (102). In suggesting that we ourselves may in essence be "modes of a single field, existing through participation," Pesic recalls not only the (w)holistic superself of Spinoza,[22] *The Thin Red Line* (1998), and many non-Western cosmologies in general, but—from a different point of observation—Heidegger's *Mitsein* and Nancy's "being singular plural."

Furthermore, the pursuit of quantum computing has highlighted the possibility of inhabiting a "superposition," of being an entity that exists in two places and in two different states at exactly the same time. This most radical manifestation of the both/and hypothesis goes beyond those conflicts which arise from being both a mother *and* a lawyer, or being gay *and* homophobic. It is a literal ontological rendering of something which we have yet to develop the conceptual paradigms for, at least outside the specialized vocabulary of postclassical physics.[23] What happens, then, to the all-pervasive Platonic legacy when 0 and 1 can be *both 0 and 1 at the same time and in the same place?* Is this the end point of the grand narrative? Is this the moment the stars wink out one by one, since the universe has successfully enlisted humans to solve its own paradoxical "issues" with fusion and difference?

One suspects not, if only because this would be to confuse the universe's issues with our own—once again to conflate That-Which-Is with all those humble methods we employ to *account* for That-Which-Is.[24]

Thus, we see how contemporary science can in fact deconstruct the hallowed tenets of classical science, and thus the foundations of liberal ethics and ontology.[25] As Nabokov has already told us, no matter how singular we may appear, we are in fact all anagrams of each other—despite the proverbial snowflakes, or the fingerprints on our hands, which seek both similarity and difference in the hands of another.

But perhaps this is too easily recuperated into abstract and vacuous universalism. Rather, we should say that we are *potentially* and *partially* anagrams of each other. Tom may be an anagram of Dick but not an anagram of Harriet, whereas Harriet may be an anagram of someone else (who, paradoxically enough, is an anagram of Tom).[26]

Yet, (sub)cultural and sexual difference complicates the "grammar of belonging," or the concept of ontological anagrams, since we read differently depending on our upbringing: right to left, left to right, against the grain, between the lines. If we are "spoken" by language, as everyone from Heidegger to Hamid Naficy agree, then ontology is essentially cultural.

As Nancy writes:

> It is, each time, the punctuality of a "with" that establishes a certain origin of meaning and connects it to an infinity of other possible origins. Therefore, it is, at one and the same time, infra-/intraindividual and transindividual, and always the two together. The individual is an intersection of singularities, the discrete exposition of their simultaneity, an exposition that is both discrete and transitory. (2000, 85)

That new metadiscipline known as cultural studies has been sending itself to sleep with mantras based on the notion of "self-reflexivity." However, this well-meaning concept is usually based on an altogether too reified notion of the self. Rather, we should begin with reflexivity, itself creating (and constantly recreating) the self from external zones of indeterminacy. In this way we avoid treating the I as the original source of ethical feedback, which is seen as channeling back toward a stable, preestablished, suspiciously coherent entity.[27]

"Aircraft are nothing more than the visible components of an integrated international air transport system," writes Pierre Lévy (1998, 112). But couldn't we say the same about individuals within a libidinal economy? The challenge is to step back from the table in order to see the human in holographic (rather than holistic) terms, to see the potential ways in which each piece contains an inscription of the whole.

The reader may quite rightly ask, Why do this? Why insist so persistently—often against the dictates of consistency or decorum—that we reconfigure our metaphysics? Why, at the risk of sounding like a flaky New Age prophet, do I push the Leibniz-lite position that monads are "perpetual mirrors of the universe"?

The answer is as simple as it is difficult to actualize: We are far less likely to lock up foreigners in privatized gulags, or to fly hijacked planes into skyscrapers, if the very notion of "foreigner" is completely foreign to us.

Moreover, if we resist the historical temptation to identify and then worship a "fundament" to humanity, then there is no recourse to fundamentalism, whether it be Christian, Muslim, or liberal-humanist.[28]

Yet, we posthumans are nothing if not stubborn creatures, always insisting on the masochistic and claustrophobic comforts of the lover's discourse; of which the prime directive is that we consider ourselves unique while simultaneously (and surreptitiously) acting on the interchangeability of others.

The genuine anguish which results from this constitutive glitch in the code prompts David Odell to question "the quixotic desire for uniqueness." He writes: "This desire is incoherent and groundless . . . yet preserves itself from that knowledge by a general clouding up of the mind. One cannot be unique in knowledge anyway, the closer affinity is with pain which seems the prime datum of apartness." Linking this particular configuration of pain and perceived solitude, Odell goes on to state: "Metonymy, or especially synecdoche, has always appeared to me the most sexual of tropes. And conversely, sexual desire itself seems to be a performative synecdoche. Romance is thus metonymy pretending to be metaphor" (2001). In other words, the desire to select one example of the human race and elevate him or her to the exceptional—*yet representative*—status of all examples, is a disingenuous desire. As with all examples, the exemplar is simultaneously and paradoxically the representative and the exception. And yet, the pleasures of throwing a halo around the beloved stem from the disavowal of this very situation (i.e., the necessary conditions of its own emergence).

The remarkable thing about love, however (and the secret of its virulent resilience), is its capacity to flood the system of the lover with a toxin that creates hostile indifference toward the contradictions inherent in its own compulsions and rhetoric. As Robert Musil observes:

> There are traces of this in even the most commonplace situations of love: the charm of every change of clothing, every disguise, the meaning two people find in what they have in common, the way they see themselves repeated in the other. This little magic is always the same, whether one's seeing an elegant lady naked for the first time or a naked girl formally dressed for the first time in a dress buttoned up to the neck. And great reckless passions all have something to do with the fact that everyone thinks it's his own secret self peering out at him from behind the curtains of a stranger's eyes. (1996)

Hence the strategies of artifice and evasion that certain subcultures have developed in order to avoid these "viscous circles" of narcissistic love.[29]

For instance, being "camp" is one method of adopting a transdividual, generic personality, thereby relieving the social pressure to display personal idiosyncrasies. Indeed, it is possible that being "flamboyantly gay" is a statement against the hypocritical injunction to be unique, to be "yourself" (hypocritical since anyone who sincerely accepts the invitation to be either unique or "yourself" is vulnerable to being locked away in a mental institution). Moreover, it is possible that this strategy also works with *all* public masks—the intellectual, the activist, the poet—all of which come with a set of operating instructions and built-in responses to various situations and environments.

Plato's mouthpiece, Pausanius, claims that "the actions of a lover have a grace which ennobles them; and custom has decided that they are highly commendable and that there is no loss of character in them" (1999, 711). What we have been encouraging, however, *is* this very loss of character. For while the sloganeers at Vodafone are not wrong in insisting that "how you feel depends on who you're connected to," they do not go quite far enough, since it is more the case that "*who you are* depends on who you're connected to."

And so, we return again to the trope of the handshake. Indeed, we do well to remind ourselves that *handshake* is also the name given to a modem's process of connecting to another modem, which creates that strange squealing sound on dial-up systems. This technological connection introduces the modems to each other so they can establish the transmission speed, whether they will employ error correction or compression, and other agreements about how they will exchange information.

On October 29, 2002, scientists in Britain and the United States initiated the "first transatlantic handshake over the Internet." Reuters reported:

> In a technological first, they will use pencil-like devices called phantoms to recreate the sense of touch across the Atlantic. . . . The phantoms send small impulses at very high frequencies down the Internet using newly developed fiber optic cables and extremely high bandwidths. When a scientist in London prods a screen with the phantom, the sensation should be felt by a colleague in Boston, and vice versa. "Pushing on the pen sends data representing forces through the Internet that can be interpreted by a phantom and therefore felt on the other

> end," said Mel Slater, Professor of Computer Science at University College London (UCL). "You can not only feel the resulting force, but you can also get a sense of the quality of the object you're feeling—whether it's soft or hard, woodlike or fleshy." (2002)

Subsequent reports suggested that the experiment was so "real" that the participants had bruises on their hands after a vigorous session of tele-introductions. This indeed brings us full circle back to Stiegler's primal handshake metaphor and the transductive relationship that it underscores. For as Kenneth Gergen insists: "We can replace the Cartesian dictum *cogito ergo sum* with *communicamus ergo sum*, for without coordinated acts of communication, there is simply no 'I' to be articulated" (1991, 242). At the same time, however, Gergen acknowledges, "If I extend my hand and smile, the gesture hovers at the edge of absurdity until reciprocated by another" (242). Shame, once again, enters the equation—or at least a sense of self-consciousness that helps delineate (or even produce) the self-which-is-feeling-self-conscious.

Reading these interconnected phenomena through a diffuse "we," Agamben hits the nail on the head when he writes that

> the planetary petty bourgeoisie is probably the form in which humanity is moving toward its own destruction. But this also means that the petty bourgeoisie represents an opportunity unheard of in the history of humanity that it must at all costs not let slip away. Because if instead of continuing to search for a proper identity in the already improper and senseless form of individuality, humans were to succeed in belonging to this impropriety as such, in making of the proper being-thus not an identity and an individual property but a singularity without identity, a common and absolutely exposed singularity—if humans could, that is, not be-thus in this or that particular biography, but be only *the* thus, their singular exteriority and their face, then they would for the first time enter into a community without presuppositions and without subjects, into a communication without the incommunicable.
>
> Selecting in the new planetary humanity those characteristics that allow for its survival, removing the thin diaphragm that separates bad mediatized advertising from the perfect exteriority that communicates only itself—this is the political task of our generation. (1993a, 65)

Conclusion

Of Mice and Multitudes

> The true situation is that there is not enough technics; it is still very rough. The reign of capital curbs and simplifies technics, whose possibilities [*virtualitiés*] are infinite.
>
> —ALAIN BADIOU

We have covered a lot of territory since the opening pages of this book.

From the outset, we noted that technology is, above all, a set of relations. We then established, even more strikingly, that technology underpins the *will* to relations themselves (even if such a will is initiated for self-seeking purposes). Those discourses which support, mediate, and transmit "love" and "community" were then *re*-presented as technological forms—not in the instrumentalist sense of *Technik*, but rather in the spirit of *techne* and *poiesis*, as a mode of revealing or "bringing forth."

But the question remains: What exactly is brought forth by this triangular dialectic? Typically, the answer is another question, *the question concerning technology itself*—that is, the ongoing challenge to address political questions of sovereignty, power, autonomy, difference, and plurality through an ontological lens; to understand that the daily discourse of belonging is produced by an anthropological machine that constantly calculates who is human and who is not, what is of value and what is not, what matters and what does not. As such, this machine seems programmed to negate the very possibility of people gathering under the blank flag of what Agamben calls "the coming community"—a concept designed to confuse the functioning of this machine's logic, rather than as a blueprint for yet another predoomed utopia.

Our inquiry then focused on the privileged position of love within the assemblage and asked whether the sociohistorical "codification of intimacy" helps or hinders Agamben's vision of an "inessential commonality." The answer, predictably enough, is that it does both, depending on the context. For on the one hand, love locks us into an individualistic and egotistic mode, reiterating and reinforcing the dogma of neo-Christian liberal-humanism, with all its associated baggage (the same baggage we take with us to the analyst's couch, the confession booth, the liquor cabinet, the medicine cabinet, the talk show circuit, the Self-Help section, the dessert trolley, the goat cheese diet, the StairMaster® machine, and even Lover's Leap). On the other hand, once freed into a more nuanced understanding of "essence" as outside the solitary confinement of the self, then the lover's discourse has the potential to neutralize the severely compromised mantras of modernity. That is to say, Agamben's notion of whateverbeing, or Nancy's notion of being singular plural, illuminates the exit light from the current libidinal economy, whereby difference is measured in narcissistic units according to its distance from the quarantined, paranoid self—a self constantly under siege from external forces and abstract otherness.

We have encountered many figures who could potentially qualify as constituents of the coming community, as noncitizens of an organless body politic. One of the "common" features paradoxically shared by these figures "who have nothing in common" is their refusal to identify themselves according to the overdetermined criteria of the state or the market (or indeed any interpellative force, such as the various demands of jealous lovers).

And so, in these concluding pages, it would no doubt be both useful and appropriate to anticipate some of the criticisms that could be directed against both my argument and the changes in orientation that it necessitates.

First of all, isn't my target something of a straw man? Isn't this so-called Cartesian-Hegelian-Kantian subject a special effect of those endless metacommentaries and secondary sources, which then use such a caricature for their own propaganda? Well, quite simply, yes—and here I am no exception. But the very fact that this straw man stalks the psychosocial landscape of the West, like the scarecrow in *The Wizard of Oz* (who is, for all intents and purposes, quite brainless), makes my case no less valid. Plato, Descartes, Hegel, and Kant were, of course, never as straightforward or naïve as their detractors depict; however, the *simplification* of their system continues to soak into the fabric of everyday life. And as a result, this straw man is remarkably resilient to critique, precisely because he is *not* flesh and blood. The myth of the unified individual goes on, despite the fact that the very fiber of his being is composed entirely of "last straws" (including the one that broke Nietzsche's camel's back).

Thus, I have indeed been attacking a straw man, a wicker man, a man with too many qualities; for this man harnesses other men to speak for him, to pull us by the ear when our answers are "wrong" or slow or flippant. This man, like the citizen of the modern state, is simultaneously everyone and no one. It is thus our tasks both to appreciate the complex ways in which different discourses intersect, interact, and mutate; while at the same time to acknowledge the ways these same discourses congeal into stable, entropic (even caricatured) forms, so that they may be more easily transmitted to the highest number of people.

The second preemptive critique is raised by Kevin Robins, who writes, "It is the continuity of grounded identity that underpins and underwrites moral obligation and commitment" (2000, 85). If this is the whole truth, then Agamben's project will necessarily fail. Robins's position, however, contains a sliver of reactionary nostalgia for a solid, "grounded" identity and does little to counter more rigorous and sophisticated modes of post-subjectivity than his favorite (and deserved) target, politically naïve techno-utopianism. Robins of course is not the only person to believe that one must first have a strong identity in order to sympathize with others. However, in these rhizomatic, postmodern times, such calls for the grounding of identity seem as hollow in motivation as they are sober in tone.[1] Agamben is correct in pointing out the manifold ways in which both the state and the Spectacle can co-opt any claim based on identity, whether it be made on the level of sexuality, citizenship, or . . . *whatever*. As Deleuze and Guattari so forcefully argue, European racism "has never operated by exclusion, or by the designation of someone as Other." Rather, it "propagates waves of sameness until those who resist identification have been wiped out" (1999, 178). Oppositional (identity) politics will always be locked into the same logic as the force it opposes. What is required then, is an *a-relational* politics.

The third possible critique is launched from a very similar position, namely the rather pointless investment in either chickens or eggs, such as Ulrich Beck's call for a "subject-orientated sociology" (Beck and Beck-Gersheim 1996). Beck argues that contemporary paradigms fail to take into account the effects of individual decision making and action. However, such "returns to the subject" represent an anthropological-humanist throwback to questions of "agency," smuggling back many of the metaphysical barnacles that insightful thinkers have managed to scrape off the Ark over the last century. It also stands as one example, of many, of Pynchon's dictum "If they can get you asking the wrong questions, then they don't have to worry about the answers."

A fourth possible critique relates to the question of conflation—specifically, my fondness for splicing together discourses as seemingly disparate as love, technology, and community. After all, is not the wisdom of scholars based on their ability to make distinctions? To dissect, typologize, and taxonomize? In my defense, however, we have witnessed the technical skill required by Hephaestus to make one back out of two. And more importantly, such conflations are done precisely (and I readily admit that the term "precisely" is overused in current European philosophy, possibly as a compensation mechanism, since precision is often exchanged for rhetorical force), *precisely* to avoid the notion that "we are all one."[2]

To approach the problem of belonging techtonically is not, one would hope, to indulge in semimystical hippie talk. It is to tie knots where the connections were too smooth or the differences too exaggerated. At the same time, it is to undo deep-seated assumptions about the conceptual apartheid of subjectivities. It is to suggest that Agamben's coming community, Blanchot's unavowable community, and Nancy's inoperative community are among the best guides we have for navigating the dangers associated with this new, already-exhausted century.[3] My dependency on "qualified conflation" thus reflects this prescient point of Heidegger's: "Every nationalism is metaphysically an anthropologism, and as such subjectivism. Nationalism is not overcome through mere internationalism; it is rather expanded and elevated thereby into a system. Nationalism is as little brought and raised to humanitas by internationalism as individualism is by an ahistorical collectivism" (1998, 260).[4]

Finally, there is the possible critique concerning the historical motivation of such anti-identity projects—specifically, the seemingly suspicious timing that prompts "white Western heterosexual males" to deconstruct the modern subject at the very moment that those who historically have been excluded (women, people of color, noncitizens, the poor, homosexuals, et al.) are claiming access to legitimate identities within the system. However, this critique fails to take into account the overdetermining role of this *very same* system. From a certain perspective, it is indeed heartening that black women (to take merely the most tokenistic example) are beginning to become more visible in the public eye. However, if we pause to think that such figures include Oprah Winfrey and Condoleezza Rice, then we find little to praise in such individual instances of "empowerment." Skin color, even personal history ("I came from the ghetto") means next to nothing when it leaves the mechanisms of exclusion and injustice in place—indeed, when it only encourages them further (in a Bill Cosby effect) to subscribe to the same disingenuous mantras of "hard work," "human dignity," and insidious assimilationism.

And thus the challenge is not to maximize the opportunities of the marginalized to join the gluttonous feast at the Captain's Table but to question the logics of identity which allow such traditionalist rankings and lopsided distribution in the first place. Again, the charge may be made that this is not very pragmatic; that we need to "address material issues on the ground." However, the United Nations, as much as the White House, exploits this notion of pragmatism in order to implement the (fundamentally antidemocratic) agenda of capital and to block any critique. Indeed, one need only be sensitive to the ways in which they do this, in order to suspect the division of "materiality" and "ideology" which sustains, in turn, naïve critiques of the most important project bequeathed to those with an interest in political justice—that being the deconstruction of the self-congratulatory liberal-humanist subject.

As long as we address only symptoms, then we will become dizzy and disoriented from tracing an endless vicious cycle. Of course, this does not mean, for instance, that we should be against facilitating debt relief to Africa. Such Band-Aid pseudo solutions, however, should be stopgap measures only, while we concentrate on the far more endemic task of dismantling the structures that inevitably produce famine in so-called developing countries in the first place. This is quite a simple plea, you would think, but it is one that seems to be falling on deaf ears, as much in "cultural left" circles (who have no truck with this peripheral "philosophy" stuff) as the IMF and the World Bank.[5]

To be blunt: Ontology addresses who and what exists, under what regimes of meaning, and for whom. It is thus intrinsically *political*, with direct, material effects and implications. And so there is no valid reason why the "global justice movement" need fear the subtleties of so-called Theory. In fact, these two camps, both exasperated with the status quo, can sustain each other in creating the conditions for an emergence on the other side of exhaustion.[6]

But as long as we insist on listening to self-empowerment training cassettes and shouting, "You go, girl!" very little will change in this individualist-isolationist economy of divide and conquer. That is to say, the only way to defy the sinister logic of "Every man for himself" is to erase the category of "man" which comprises its very foundation.

Re: <no subject>

> Humankind fails itself in a certain sense and has to appropriate this failing.
>
> —GIORGIO AGAMBEN, *The Coming Community*

> Frankly, I expected better from Jimmy the Scumbag.
>
> —CHIEF WIGGUM in *The Simpsons*

Of course, it is one thing to realize these things on a conceptual level, but how are we to live the admittedly poetic possibilities of the coming community? The siren song of consumer culture (along with its democratic shadow) is so powerful and all-encompassing that we are always already jolted back into the idea of the atomistic individual. Furthermore, it is not simply a struggle between whateverbeing and the Market Spectacle, since each manages so deviously to hijack the other for its own purposes. Nevertheless, there *is* a struggle going on, and this is between different assemblages of power, economics, rhetorics, and knowledges. At stake in this struggle is a metaculture trying to get past the "post" of the human; to see the human as somehow both belated and ahead of itself—as arriving too late for the community which hasn't yet arrived.

As with Alice's Wonderland, the logic of this place—*our* place, today—is both bewildering and punctuated by seemingly arbitrary, illogical violence ("Off with his head"). The confusion that results, however, holds within it enough potential energy to realize—and more to the point, *continue* realizing—that things could be otherwise.

If we wish to talk in the vulgar terms of "mistakes," then two of the most significant occurred when we decided to place humans at the center of the universe, and then the self at the center of humanity. This decision has solidified into subconscious habit, even after Copernicus (perhaps as a vast compensation mechanism). It is difficult to disagree with the fact that what we call "life," however, is one massive cosmic recycling program. Moreover, it is this very programmatic aspect that some more adventurous thinkers call "intelligence" or "essence." The human effort to establish uniqueness and ontological sovereignty is the resistant factor. It consists of the futile and vindictive struggle against the greater pattern, against greater nature, against each other—that is, against "time," which incorporates the virtual as much as the actual. (This is not to advocate a mystical fatalism but somehow to constructively embrace this knowledge, somewhere in the middle of the bridge that spans Nietzsche's will to power and Blanchot's radical passivity.)

Since we are composed entirely of elements that precede "us," physically and metaphysically, our essence is external. It comes from elsewhere. Each of us has only *leased* our "self" from the cosmos and must one day give it back again.[7] Thus, we should realize the time has come to renounce our self-designated diva status and admit that we humans are not the star of the story. The *story* is the star of the story, and the sooner we assimilate this understanding into our thoughts and our actions, the more interesting the story will become to us—not only that, but the more interesting we will become to all those other characters we condemn to supporting roles (animals, artifacts, machines, and so on).[8]

To love someone is to love his or her particular pattern. Thus, the sphere of love is a matter of aesthetics ("I don't know much about art, but I know what I like"). The stupid attitude toward the miraculous indifference of the vortex (as opposed to the void) is nihilism. Here nihilism is understood as the hostile *ressentiment* toward the pattern, which usually prompts the attempt to impose "one's own" pattern on it, as happens with all fundamentalisms, including purist Apollonian art. This is why Nietzsche called Christians "nihilists": because of their quaint anthrocentric daydreams of a life after death in which we retain our former identities like a security blanket. Aesthetics *is* ethics, in that to love a piece of the pattern is to metonymically—holographically—love the whole pattern, rather than to isolate one piece, frame it, and hang it in the museum of the heart.

Regrettably, *God* is the name most often bestowed on the pattern itself—that is, the techtonic sounds and movement of shifting sands, migrating birds, rolling waves, beached whales, chatting printers, whistling modems, moaning lovers, and collapsing buildings. As Nietzsche quite rightly insisted, to name something is to foreclose alternative access to that which is named. That is to say, a God by any other name could smell far sweeter.

And since we are wrapping up this journey with the theme of mistakes, it is certainly a grave error to believe that the United States enjoys "technological superiority" over, say, Vietnam, Afghanistan, or Iraq. No one refutes this claim, although many point out that this so-called superiority has limits when confronted by the "stubborn" or "passionate" will of those people the Americans happen to be fighting. This stubbornness, however, is *itself* a technology. Islam, Christianity, communism, ethnicity, patriotism, nationalism, and communalism are all techniques of relation, and in a very instrumental and precise sense. These so-called ideologies are in fact self-calibrating cultural codes which can mobilize people and passions: love, hate, ignorance, violence.[9] Tanks and smart bombs are merely the condensations or excreta of these less visible, but no less deadly, technologies. They all form the same matrix, which can conduct sympathy and torture through the same circuit, if need be.

This observation can be extended over the globe. Anthrax occurs in "nature," but when it is engineered into fine particles and sent through the mail, it has become a human-inflected technology. It promotes a community of the infected. We thus create a situation of bombs and bricolage—Afghan blacksmiths making pots and pans out of spent American missile shells, and the Vietcong recycling military scrap metal into improvised and terribly effective hand grenades.

Yet once again we must, for the sake of methodological consistency, ask the corny question "Where is love?" Well, in a word, everywhere. Love for one's brother, mother, country, freedom, or God clears the vectors for violence against others. Sentimentality provides the oil for the world's great war machines, just as those romantic hearts, which flutter with love, pump the same blood that is only too easily spilled onto the ground.

RAQ: Rarely Asked Questions

> When our identity is in danger, we feel certain that we have a mandate for war. The old image must be recovered at any cost. But, as in the case of "referred pain," the symptom against which we lash out may quite likely be caused by something about which we know nothing.
>
> —MARSHALL MCLUHAN AND QUENTIN FIORE, *War and Peace in the Global Village*

> We take an attentive interest in the life and being of many things by not treating ourselves as fixed, stable, *single* individuals.
>
> —FRIEDRICH NIETZSCHE, *Human, All Too Human*

As a concluding image, picture, if you will, a digital, animated map, one which shows not only the continents breaking apart and moving across the face of the planet, but the political borders waxing and waning. We can see the early traces of ancient Egyptian boundaries and the great northern splash of the Roman Empire, like a starburst eventually retreating into a red dwarf. The borders of medieval Europe writhe like worms across the continent, sometimes tangled and then separating into clear lines, which themselves begin floating back and forth. The Americas begin to self-partition, cracks appearing like stress fractures in cement, stretching and shrinking into semirecognizable shapes. The vast expanse of the ocean is sliced into seven seas and, eventually, national moats known as "designated fishing areas." Then comes the scramble for Africa, carved up like a dry wedding cake, followed on the other side of the globe by the blood vessels of postrevolutionary Asia and post-Soviet Eastern Europe. Even Antarctica begins to crack, not only through the effects of global warming but by the colonization of its ice cap by committees and lobby groups, scientists and governments. To these geopolitical lines we could add lines of longitude and latitude, datelines, ley lines, songlines, trade routes, no-fly zones, the Bermuda triangle, and the vicious vectors of human trafficking.

Such are the techtonic movements of our ever-evolving world picture.

Then how to respond to the succinct cultural diagnosis of one of my students: "Same shit—different century"? Are we misreading the pulse of the issue when we insist that "new media" is really something new—or indeed, that any cultural manifestation has an unprecedented impact on those "ambient conditions . . . what others who come afterward will call your culture" (Lacan 1999, 54)? Is it true, moreover (as this same student insists), that we feel the same anxiety when our email server crashes as when the medieval maiden watched bandits intercept her message-laden carrier pigeon with a crossbow?

Well, as usual, I will sidestep a direct confrontation with such questions in order to reveal the incentive behind their own momentum. This is because the answers given, whether one chooses to emphasize novelty or continuity, are not nearly as important as the reasons for doing so. Heidegger, for instance, insisted that there is indeed something "which is new" in modern technology: namely, that "monstrous" and "unreasonable" demand we ourselves make of nature, and which boomerangs back at us via the material manifestations of standing reserve. He did this because he wanted to inspire a change in our thinking, away from calculative reason and toward a more authentic understanding of Being.

And while we may not have a time machine, we do have the textual traces of former epochs to afford access to Lacan's "ambient conditions." The more we listen to such traces, the more sensitive we become to the fact that saying "Same shit—different century" forecloses any revealing discussion of the significant properties which emerge at any given epistemic shift. (Whoever has experienced three decades of life is less likely to agree with a claim that things stay the same, just as Nietzsche insisted that "the brevity of human life misleads us into making many erroneous assertions about the characteristics of human beings" [1995, 50–51]).[10] This is to say that "things" may continue to smell bad; however, the source of the stench could well be different shit for a different century—the result, perhaps, of genetically modified food or the miniscule nanobots which will soon be cruising through our veins in search of designated pollutants.

Indeed, as we enter the "nanopolitical" era, the possibilities for whateverbeing multiply, as do the dangers which it must confront. While the burgeoning biotech industry teaches us that it is foolish to assume that nature and technology are incompatible, the equally industrious quest for cloning is, at this point, creating far more questions than answers. (Surely it is more difficult to convince someone that we are experiencing "the same shit, only in a different century" when the person you are trying to convince is an exact replica of yourself.) As William Burroughs observes, cloning may enable us to finally jettison "the tiresome concept of personal immortality . . . predicated on the illusion of some unchangeable precious essence that is greedy old MEEEEEE forever. . . . The illusion of a separate inviolable identity limits your perceptions and confines you in time" (in Shaviro).[11]

Whether genetically modified crops, or cloning, or any of the other technological "advances" of our species result in monstrous angels or angelic monsters—or something we could never have anticipated—remains to be seen. That prophet of the new-media age Marshall McLuhan was at times wary of, other times ecstatic at, the possibilities offered by such "extensions" and "amputations" of the exoskeletal human nervous system. For McLuhan, "all media or technologies, languages as much as weaponry, create new environments or habitats, which become the milieux for new species or technologies." Thus, as posthumans with the ability to finally re-comprehend the world as a "total field," we should look forward to living in the "matrix" of a superterrestrial macrocosmic connubium (McLuhan and Fiore 1997, 190). Given the popular sympathy which greeted the film *The Matrix* (1999), it seems that most of us are less enthusiastic about the continuing symbiosis between humans and technology. Within such a context, the call for more responsible applications of (apparently neutral) technology necessarily falls short; primarily because such calls fail to understand that humans are, in essence, technological.

"Humanity makes pitiless use of every individual as material for heating its great machines," Nietzsche presciently wrote, "but what are the machines for, if all the individuals (that is, humanity) serve only to keep them going?" He goes on to ask whether the final irony of the "human comedy" is in fact "machines that are their own purpose" (1995, 280).

Ultimately, it is this knowledge concerning the immanent isomorphism of ourselves and our environment (or ourselves *as* our environment, and vice versa) that prompted thinkers as diverse as Nietzsche and Plato to agree on at least one thing—namely:

Nothing human is worth taking very seriously; nevertheless . . .

NOTES

PREFACE

1. The standard reference point here is Sartre's phenomenological gap between the Self-for-the-Self and the Self-for-Others.

2. For more on the ideological repercussions of such "symbolic efficiency," see Zizek, *The Plague of Fantasies* (110).

3. The significance of puns such as Nick Hornby's *High Fidelity* should likewise not be underestimated. This is especially so since Friedrich Kittler has gone to such lengths to underline the sociotechnical foundations for such "fidelity" between the reproduction of sound (the Mother's voice, His Master's voice, Goethe's voice, etc.) and the primary discourse network of love's demand to "remain faithful."

4. Bill Nichols (1996) has noted that new-media discourse tends to fetishize the equipment used to represent or access libidinal material as much as the libidinal material itself. However, this is hardly a new gesture, since those countless poems to "Cupid-bow lips" similarly fetishize the medium of love as much as the message.

5. Indeed, Freud emphasizes the "primordial ambivalence of feeling toward the father" as just such an originary love/hate nexus. "His sons hated him, but they loved him, too. After their hatred had been satisfied by their act of aggression [i.e., parricide], their love came to the fore in their remorse for the deed. . . . Now, I think we can at last grasp two things perfectly clearly: *the part played by love in the origin of consciousness* and the fatal inevitability of the sense of guilt" (1953–1974, voI.13: 143; cf. 21: 132; my emphasis). The following pages, however, attempt to extricate themselves from such a primal scene, that is, from being circumscribed by the genealogy of morals in relation to abstract community.

6. We could say of technology what Paul de Man says about Romanticism, that it embodies "the movement that challenges the genetic principle that necessarily underlies all historiography" (82).

7. Another analogy for the the primordial event of *techne*, figured via periodicity, could be made with Lurianic Kabbalah, as described by Karen Armstrong: "The primordial event described in myth [i.e., God's self-contraction, the shattering of the vessels unable to contain His light, and the unbearable exile of light in a world of evil] is not simply an incident that happened once in the remote past; it is also an occurrence that happens all the time. We have no concept or word for such an event, because our rational society thinks of time in a strictly chronological way" (11). The trick is to grapple with this type of temporality without any theological resonance. (And thanks to Justin Clemens for this particular parallel.)

8. Here we could contrast Virilio's notion of speed with Deleuze's particular rendering of "slowness." Jean Baudrillard also seems to be groping toward an understanding of techtonics when he writes that "the catastrophic form congenital to the era of simulation . . . [is] the seismic form, where the ground is missing, that of fault and failure, dehiscence and fractal objects, where immense plates, entire layers slide one under the other and produce intense surface tremors" (FS, 20). See also Robert Frodeman's *Geo-Logic: Breaking Ground Between Philosophy and the Earth Sciences* and Manuel de Landa's *A Thousand Years of Non-Linear History*.

INTRODUCTION

1. Originally titled *Lola Rennt*.

2. Of course, Manni should simultaneously realize the he is "one of those other boys." As Goethe once wrote: "If I love you, what does that concern you?"

3. Such distinctions become even fuzzier on the microbiological level, for as my G.P. has informed me, the amount of "foreign cells" living in symbiotic union with our person outnumbers "host cells" by a ratio of ten to one (thereby putting any such distinction in question).

4. In *The Dying Animal*, one of Philip Roth's characters notes that after a certain age you stop meeting new people, anyway: "Who are the new people when you *do* meet them? They're the same old people in masks. There's nothing new about them at all. They're *people*" (2001, 107).

5. Blanchot refers to these deconstructive formulas "in accordance with some incongruous words that I remember having read somewhere, not without irritation, where reference is made to the coming of what does not come, of what would come without an arrival" (Cadava, Connor, et al. 1991, 59).

6. Nietzsche utilizes a different metaphor for a similar idea, namely the person without melody: "such natures, who *become* nothing, without our being able to say that they *are* nothing" (1995, 294). In this book I am also fusing the term "whatever being" into "whateverbeing" for the sake of promoting a coherent concept and entity, rather than from a Neoplatonic tendency toward fusion.

7. See also Victoria Nelson's *The Secret Life of Puppets* (2001).

8. Significantly, the puppeteer character in *Being John Malkovich* obsessively returns to the story of Heloise and Abelard, one of the primal scenes for the Western love story. See also Giorgio Agamben's first book, *Stanzas* (1993b), for an inspired critique of "self-centered" theories of the Narcissus myth, another key quilting point for the erotic code itself.

9. For more on this syndrome, see Chris Cunningham's music videos for Aphex Twin's "Come to Daddy" (1998) and "Windowlicker" (1999).

1. LOVE AND OTHER TECHNOLOGIES

1. Friedrich Kittler goes even further, stating, "If media are anthropological a prioris, then humans cannot have invented language; rather, they must have evolved as its pets, victims, or subjects" (1999, 109).

2. For the purposes of ontology, it seems impossible to work with a definition of technology that is overly inclusive. As far as I am concerned, bread is a technology. So are carrier pigeons, blackmail, and seduction. Let us recall that Lacan often describes *jouissance* as an "apparatus," just as Foucault discusses the "technologies of the self" and Deleuze traces the "machinics" of desire. This all-embracing definition of *technics* is unpacked throughout the following chapters.

3. For the sake of consistency I retain the gendered language of Stiegler, Rousseau, and other Continental philosophers. Later sections will deal more directly with the legacy of such sexuated terms.

4. Empathy therefore precedes sympathy, because it is impossible to feel sorry for someone if you cannot imagine being in their place. (See Rei Terada's *Feeling in Theory*, especially the section entitled "Imaginary Seductions" [31–41], for a more nuanced reading of this philosophical history.) The step from empathy to sympathy is assumed by liberal-humanistic discourses based on Hellenic ethical debates. The iconic status of Hannibal Lecter in the contemporary popular imaginary, however, has much to do with his refusal to conflate the two terms. In the eponymous novel by Thomas Harris, Hannibal claims that he understands how his victims feel, but this does not necessarily make him feel sympathy for them (2000). (Thanks to Guy Chardonnens for this reference.)

5. Interestingly, Kittler describes the pre-1936 Turing machine as "a hermaphrodite of a machine and a mere tool" (1999, 18). Furthermore, he insists on the intrinsic role of physical impairments, such as blindness and deafness, in the development of the gramophone and the typewriter (22), an etiology extending back to the crippled Hephaestus. The suggestion is not the banal one that humans need to invent instruments as prostheses, but that our physical limitations *are the very conditions* for technology to autogenerate. These "extensions of man" are but man's extension, allowing Kittler to read the evolution of technology as a "sexually closed feedback loop" indistinguishable from "the decomposition and filtering of love" (182–84).

6. Lacan enigmatically states, "There is such a thing as the One." However, this particular rendering of a unity does not refer to Aristophanes' fable, since Lacan's system discounts the possibility of perfect reunion after the unity has been split. For "when one gives rise to two (*quand un fait deux*), there is never a return. They don't revert to making one again, even if it is a new one. *Aufhebung* is one of philosophy's pretty little dreams" (1999, 86).

7. The *Oxford English Dictionary* definition of *individual* includes "One in substance or essence; forming an indivisible entity; indivisible." Interestingly, one of the earliest references to the individual is in reference to the Trinity (c1425 Found. St. Bartholomew's (E.E.T.S.) II "To the glorie of the hye and indyuyduall Trynyte"), recalling Aristophanes' phrase "the sexes were originally three in number . . . man, woman, and the union of the two" (Plato 1999, 719).

8. "For the 'hospital system': the new medicine 'without doctor or patient' that singles out potential sick people and subjects at risk, which in no way attests to individuation—as they say—but substitutes for the individual or numerical body the code of a 'dividual' material to be controlled" (Deleuze 1992, 7). See also "Thought and Cinema" in *Cinema 2: The Time-Image* (2003, 162–64).

9. The fact that couples are expected to "work" at their marriage betrays a form of "romantic labor" at the heart of the code. Indeed, couples even have to be diligent when "spicing things up" via a third term, for *ménage à trois* can literally translate as "housework for three."

10. If readers remain skeptical concerning this lack of uniqueness from one person to the next, then perhaps they should consider the reasons sex workers often use the generic term "John" for all of their many-hued customers. The same could be said for "John Doe," "Jane Doe," "Petar Petrovic," "Juan de la Cruz," and even "Allen Smithee."

11. This "trauma" of the second love may not be explicitly *experienced* as a trauma but more as a "glitch in the matrix." The power of the code thus replenishes itself through our capacity to disavow this contradiction immanent to the code itself.

12. Luhmann writes, "Individuals must be in a position to receive positive feedback not only on what they themselves are, but also on what they themselves see" (16–17).

13. There is a danger here, less in terms of Luhmann's discourse than my own, of succumbing to a latent quasi-Christian form of moralizing, when automatically denouncing "ego-centrism." From a certain ontological angle, ego-centrism looks less like a megalomaniacal form of solipsism than the desire to maximize proximity and exposure to otherness ("I want to become famous so I can meet lots more people"). Although egocentrism is usually figured as a centripetal force, toward the central ego, there is always the danger—the possibility?—of turning inside out, of succumbing to the pressure of alterity and exploding into this vortex. (Just as for Girard, snobbism is an effect of mediated otherness.) Egocentrism is thus only a "problem" when it is supplemented by

a "centrocentrism" which absorbs and neutralizes otherness (cf. colonialism, Cassanovaism, etc.).

14. Of course, Luhmann does not underestimate the importance of gendered discourse, so far as admitting, "The differences between the sexes—which were emphasized in all love codes in the past, and around which asymmetries were constructed and enhanced—are toned down. . . . [T]he question today is what to do with the remains of a difference that cannot be legitimated?" (160).

15. Kittler also reminds us that in both ancient and modern times, "reading means having intercourse with the dead" (1990, 166)—a clear convergence of love, technology, and community. See also his discussion in the same book of a "hermeneutic-erotic circle" which "regulates both reading and love" (126–30) and thus links the author's pen with the reader's eyes, the lover's loins with the beloved's mind, and the mother's mouth ("the original maternal cultivator" [51]) with the son's ear, all in a discourse network which completely conflates the technics of literature with the machinery of love ("an alphabetization in the flesh" [112]). Incidentally, while we are on the subject, Kittler's primal figure of the Mother's Mouth or Voice has seemingly been replaced, or at least supplemented by, the Children's Television Workshop.

16. Strangely, Luhmann makes a passing reference to "the code's final version," in which "everyone leads a copied existence, this being the precondition for one to be able to enjoy passionate love and make it one's own" (1998, 47). Why he believes this code has a *final* version is anyone's guess (perhaps the translator's), considering his profound understanding of historical process and progress.

17. The seductive, all-embracing power of the discourse itself is resistant to Laura Kipnis's entertaining critique of it in *Against Love* (2003). For while Kipnis writes against moralistic, couple-centric incarnations of love, she encourages unexpected passion (which is merely the other side of the coin). More accurately, she is against the institutions which stifle "real" love.

18. The entire cathedral of psychoanalysis is based on absorbing the shock of this trauma and its related effects. The move from the breast to the battlefield of Oedipal domesticity then becomes a self-fulfilling prophecy machine which generates a restricted formula for desire and its interpretation. As Luhmann notes, "The lovers' conduct is no longer patterned on novels, but rather on psychotherapy" (1998, 174). Moreover, "The influence of therapists on morals (and of morals on the therapists) is difficult to estimate, but it is surely to be feared" (166).

19. Bataille: "Movement is a figure of love, incapable of stopping at a particular being, and rapidly passing from one to another" (1985, 7).

2. THE STORABLE FUTURE AND THE STORED PAST

1. This name derives from the *Matrix*-meme, whereby a figure leaps into the air kung fu style, freezes (while "the camera-gaze" revolves around the fig-

ure), and then unfreezes into a powerful kick delivered to a hapless goon. When I first began writing this chapter, in 2001, industry experts informed me that this technology had yet to earn an official name, despite its being used for several years already—yet another sign of technology's outpacing its induction into "culture."

2. In his canonical "Theses on the Philosophy of History" (1992), Walter Benjamin notes: "A historical materialist cannot do without the notion of a present which is not a transition, but in which time stands still and has come to a stop. For this notion defines the present in which he himself is writing history. . . . Universal history has no theoretical armature. Its method is additive; it musters a mass of data to fill the homogenous, empty time. Materialistic historiography, on the other hand, is based on a constructive principle. Thinking involves not only the flow of thoughts, but their arrest as well" (254).

3. This itself prompts the question whether people can retrospectively be considered having been "photogenic" before the invention of photography.

4. Regarding the libidinal "detours and circumlocutions" that make up every biography, Kaja Silverman notes that "the word 'mother' might signify a mole at the corner of a shapely mouth, which she many years later rediscovers on the face of an eighteen-year-old boy" (2000, 42). Silverman's chapter on the "Apparatus for the Production of the Image" in *World Spectators* is, like *Lolita* itself, an interesting attempt to reconcile ontology with psychoanalysis, via affective "visual condensation," thereby complicating this claim by Bergson: "Memory [is] inseparable in practice from perception, [and thus] imports the past into the present" (1988, 73). More interesting still is the recent "literary detective story" by Michael Maar entitled *The Two Lolitas* (2005), which exposes Nabokov's possible case of "cryptoamnesia" concerning an earlier incarnation of Lolita: the eponymous preteen love interest of a tragically obsessed middle-aged man, both of whom inhabit a short story by Heinz von Eschwege published in 1916.

5. See Deleuze's *Proust and Signs* (2000, 12–13 and 151–52). Proust's paradox relates ultimately to the question of economy; since, in sharp contrast to capitalism, the more a person accumulates in life, the more he or she has simultaneously "lost."

6. The inadequacy of photography was flagged at its conception. In 1850 Nathaniel Parker Willis lamented the fact that no photograph could capture the true spirit of the celebrated Swedish singer Jenny Lind: "[N]ot even a daguerreotype was reasonably like our feeling of what a likeness should be." While Willis insisted that Lind possessed a "singular beauty," he had to admit that "the pictures of her represent the plainest of commonplace girls" (Kunhardt 1995, 88).

7. This is something which the cable channel Eurosport hopes to achieve with its concept of replaying a sporting event and calling it "relive."

8. Against the notion of linear time (in which C succeeds B, which succeeded A), we have Bergsonian duration, in which C contains B and A. In other words,

C is the contraction of both B and A. In Deleuze's terms, "if the present was not past at the same time as present, if the same moment did not coexist with itself as present *and* past, it would never pass, a new present would never come to replace this one" (2000, 58).

9. Recall also Plotinus, who believed that "identity . . . is pure being in eternal actuality; nowhere is there any future, for every then is a now; nor is there any past, for nothing there has ever ceased to be" (Pesic 2002, 145).

10. Even his name, Humbert Humbert, suggests the noncoincident character of subjectivity: a self raised to the second power.

11. One need only recall the "differences" between Nabokov and Kubrick when the former moved to Hollywood to write the script for *Lolita,* most of which was ignored by the director. Kubrick also chose to ignore Humbert's own proleptic request while looking at the wanted posters in a post office: "If you want to make a movie out of my book, have one of these faces gently melt into my own, while I look" (Nabokov 1991, 222).

12. It is, of course, obvious that "the media" has no coherent, monolithic agenda. Television in particular sees no conflict in promoting the pop pedophilia of Britney Spears and her handmaidens while simultaneously sending out the shrill condemnation of pedophiles themselves, either convicted or simply suspected. The implication is that it is somehow more tolerable to assault an adult, simply because he or she has more "experience" (i.e., has accumulated more "time").

13. A fascinatingly oblique intertextual reference to Nabokov can be found in W. G. Sebald's stunning novel *The Emigrants* (1997), in which the master himself makes a cameo appearance around both Ithaca and Montreux as "the butterfly man." This unnamed figure haunts Sebald's text, even more peripherally than in Nabokov's own works, as the elusive spirit of an émigré author wandering stealthily and knowingly from novel to novel. Likewise, the historical photographs included by Sebald serve to make this novel's "search for lost time" even more poignant than simple text would allow, according to a desexualized (but nevertheless erotic) Humbertesque logic. As such, *The Emigrants* stands as an example of lo-tech "multimedia," which draws much of its power from the incongruity, yet mutual support, of images and words. Ironically, and in contrast to much of the narrative, the final two pages contain no pictures, although this last section deals explicitly with a collection of photographs, "carefully sorted and inscribed" and "tinted with a greenish-blue or reddish-brown, of the Litzmannstadt ghetto that was established in 1940 in the Polish industrial centre of Lodz, once known as *polski Manczester*" (235–36). It is as if the uncanny and undeniable power of a photographic frozen moment only increases its affect/effect on the reader/viewer by belatedly absenting itself, leaving the ultimate image up to, appropriately enough, the imagination. The three "daughters of the night, with spindle, scissors and thread," who may be called "Roza, Luisa and Lea, or Nona, Decuma and Morta" (237), cannot be simply

fixed and represented, because they embody all those vanished lives which cannot be salvaged by Sebald and "the entire questionable business of writing" (230).

14. Humbert: "The science of nympholepsy is a precise science. Actual contact would do it in one second flat. An interspace of a millimeter would do it in ten. Let us wait" (Nabokov 1991, 129).

15. In a passage which could effortlessly be applied to Humbert's relationship to Lolita, Thomas Wall writes: "There is already something artificial about the real, something aesthetic, uncanny, plastic—if you like, something fake. Life is always very nearly a novel, an image, a corpse. Put differently, the real is always vulnerable to the stoppage of time—to the image of finitude—that it wears on its face. Insofar as this person resembles herself, she is infinitely vulnerable, infinitely fragile, as an image is fragile. A strange weakness pervades her that we cannot grasp, a bleeding we cannot staunch. She is not the same thing as her resemblance to herself, but she is nothing other than it. Human reality is nothing other than this infinite vulnerability, inequality-in-itself, or difference-in-itself" (1999, 23).

16. As Spinoza writes: "*A man is as much affected pleasurably or painfully by the image of a thing past or future as by the image of a thing present.* Proof.—So long as a man is affected by the image of anything, he will regard that thing as present, even though it be non-existent. . . . Thus the emotion of pleasure or pain is the same, whether the image be of a thing past or future . . . [since] the body is affected by the image of the thing, in the same way as if the thing were actually present" (*The Ethics*, Prop. XVIII).

17. Elsewhere Humbert laments his lustful obsession toward Lolita, admitting that he felt "as if I were sitting with the small ghost of somebody I had just killed" (Nabokov 1991, 140).

18. Gender is not so significant in terms of this melancholic-erotic structure, nor in terms of the wider love-community-technology equation, for it would be reasonably easy to transpose this reading of *Lolita* to Thomas Mann's *Death in Venice*, and to a host of other stories based on longing and preemptive loss.

19. Agamben divides this economy into three strata: the double polarity, demonic-magic, and angelic-contemplative. These correspond in our context to Quilty/Annabel, Lolita, and Literature, respectively.

20. This is not to say she *has* an essence. Just as a judge represents the essence of the Law (which is everywhere and nowhere), Lolita represents the essence of nymphancy.

21. For a parallel perspective, see Deleuze's *Difference and Repetition*, which states that "virtual objects exist only as fragments of themselves: they are found only as lost; they exist only as recovered" (1994, 102).

22. In a preemptive echo, Nietzsche states, "Only through the forgetfulness of this primitive world of metaphors, only through the invincible belief that *this* sun, *this* window, *this* table is a truth in itself, in short only because man forgets

himself as a subject and, in particular, as a subject of artistic creation, can he live in a world of repose and security" (in Agamben 1993b, 151).

23. This "purse" will resurface as the figurative herald of the literal prostitution that Lolita is forced to endure while under Humbert's "care."

24. Perhaps Plato's belief that fantasy equates to deed in the realm of lust is less a technology of moralistic self-surveillance than an acknowledgment of the ontological validity of such phantasms.

25. Nabokov's student Thomas Pynchon describes a similar scene in *Gravity's Rainbow:* "So it has gone for six years since. A daughter a year, each one about a year older, each time taking up nearly from scratch. The only continuity has been her name, and Zwölfkinder [a theme park run for and by kids], and Pökler's love—love something like the persistence of vision, for They have used it to create for him the moving image of a daughter, flashing him only these summertime frames of her, leaving it to him to build the illusion of a single child . . . what would the time scale matter, a 24th of a second or a year (no more, the engineer thought, than in a wind-tunnel, or an oscillograph whose turning drum you could speed or slow at will . . .)? (1991, 422, Pynchon's ellipses)

26. There are, in fact, forty names, since the list includes a "vagrant auditor," Aubrey McFate, who becomes an appropriately named agent in Humbert's own reading of the events which follow.

27. As is often the case, we can find classical templates for modernist conceits. In this case, Nabokov's idealized amalgam echoes the methods of the Hellenic painter Zeuxis, described as "one of the very first great masters," working during the fourth century B.C. When commissioned to paint the portrait of Helen, Zeuxis is said to have selected five beautiful women as models and blended their various features in order to produce his work. See also the 1949 Ronald Reagan movie *The Girl from Jones Beach* for a variation on the same theme. (Thanks to Rick Waswo for these examples.)

28. In *Stanzas* Agamben extends such a notion to medieval interpretations of intelligence, especially refracted through Aristotle, in which this faculty is perceived as "something unique and supraindividual—within which individual persons are simply, to use Proust's beautiful image, *co-locataires* (co-tenants, co-inhabitants), each one limited to furnishing its distinct point of view to the intelligence" (1993b, 83). As Kittler notes regarding Goethe's construction of the single idealized Woman, "The author of *Werther* granted himself 'permission to model [his] Lotte on the figures and characteristics of several pretty young girls'" (1990, 131). Even Goethe's esteemed biographer, Johann Eckermann, is sensitive enough to note that the great man, as presented in *Conversations with Goethe*, is in fact "*my* Goethe," since an inevitable "mirroring occurs, and it very rarely happens that in passing through another individual no specific characteristics will be lost and nothing foreign will be mixed in" (Damrosch, 30).

29. Similarly, a DVD player is said to either "read" a disk or "write" to a disk. But if we are watching a movie, the machine must be *simultaneously* reading and writing, or else we would see no picture. This process can be extended—or rather, *in*tended—to the act of reading a novel, in which the symbols (letters, words, sentences) are simultaneously written to the literate faculty of the mind (e.g., Freud's Mystic Writing Pad). "Pure" reading is indistinguishable from a blind silence (or a broken DVD player).

30. Arthur C. Clarke is said to have coined the following truism: "Getting information from the internet is like getting a glass of water from Niagara Falls."

31. See Agamben's *The Open: Man and Animal* (2002).

32. Regarding "the textual work that comes out of the spider's belly," Lacan writes, "It is a truly miraculous function to see, on the very surface emerging from an opaque point of this strange being, the trace of these writings taking form, in which one can grasp the limits, impasses, and dead ends that show the real acceding to the symbolic" (1999, 93).

33. While this could be seen as a step toward the embrace of a Deleuzian "nomadic science," it would, however, first have to relinquish its investment in beginning with the organism itself, no matter how extended, since the Body Without Organs has no such internal coherence.

34. In relation to our decidedly unscientific discussion of spiders, Godzich points out the discursive flows between "literature" and "biology": "Take any issue of a journal such as *Cell Biology*, and you will see on every page references to transcription, translation, and protein-folding in organisms. It is up to us now to recover these catachrestic terms from biology and to rethink what they mean in literature after their sojourn in biology" (2000, 8–9).

35. One suspects that Kaja Silverman could redeem the sentimental overtones of redemption itself, as a trope, in the novel's narrative. Thus, in finally appreciating Lolita as a *singular* reincarnation of Annabel, Humbert enters the realm of the world spectator: "We look in the way that makes appearance possible only when we also allow the perceptual present to *reincarnate* or *recorporealize* the past—to give it a *new form*. We only give the gift of Being to something when we permit it *inaccurately* to replicate what was" (2000, 145, Silverman's emphasis).

3. IN THE ARTIFICIAL GARDENS OF EDEN-OLYMPIA

1. Concerning our yet-to-be-acknowledged posthumanism, Donna Haraway anticipates that "[t]here might be a cyborg Alice taking account of these new dimensions" (1991, 154).

2. Ballard is well aware of such once-fashionable theories of hyperreality and the simulacra, as captured in the reference to a "young French waitress, who

wore jeans and a white vest printed with a quotation from Baudrillard" (2001, 88). Such theories, first popularized in the 1980s, have had something of a comeback thanks to the naïvely explicit reading of Baudrillard in *The Matrix* (1999), a text which also feels compelled to place Carroll's Wonderland at the head of its own genealogy of para-spaces.

3. In his article "Disneyland with the Death Penalty," William Gibson writes that the "sensation of trying to connect psychically with the old Singapore is rather painful, as though Disneyland's New Orleans Square had been erected on the site of the actual French Quarter, obliterating it in the process but leaving in its place a glassy simulacrum. . . . The word infrastructure takes on a new and claustrophobic resonance here; somehow it's all infrastructure" (1993). See also Andrew Ross's *Celebration Chronicles* (1999) and James Kunstler's *The Geography of Nowhere* (1993) for more on hyperdesigned, corporate-centric living spaces.

4. It seems strange that a writer as meticulous as Ballard would make a reference to "repressed executives like Alain Delage" (352) and also to their "perverse" side (387), when he has taken so much time to explain away such Freudian categories as anachronistic. However, the notion that he is trying to sneak a repressive economy in through the back door, as it were, is just as unlikely as a slip of the pen.

5. Ballard's book can therefore be read as a response to, and a revision of, aphorism twenty-seven of Guy Debord's *Society of the Spectacle*, which emphasizes "the success of separate production as production of the separate" (1977). The concept of *production* of course becomes more complex in our networked age of "knowledge economies" and "creative industries." (See Geert Lovink's public lecture, "The Principle of Notworking: Concepts in Critical Internet Culture," available online).

6. The narrator also makes reference to Polish whores from his RAF days, "scarcely women at all but furies from Aeschylus" who were disgusted by any show of feeling. "Warmth and emotion were the true depravity. They wanted to be used like appliances rented out for the hour" (Ballard 2001,156). Such memories suggest a nostalgia for a "purer," less hypocritical form of reified sexual commerce, as if this were the most we could ask for in the era of "the complete triumph of the spectacle" (Agamben 1993a, 80).

7. For a more traditionally gendered psychoanalytic interpretation of this "literary" mirror stage as well as its schizophrenic legacy, see Dana Becker's *Through the Looking Glass: Women and Borderline Personality Disorder.*

8. Concerning Lacan's use of the term *orthopedic,* Madan Sarup notes, "There is a suggestion that the identity of the subject is something added, something that helps you to stand up straight within yourself" (Sarup 1992, 65). In contrast, Jane becomes increasingly languid as she succumbs to her new environment, and is rarely depicted standing up in the second half of the book. Ballard of course never misses an opportunity to introduce erotic-prosthetic

elements to his novels; hence, the narrator becomes absorbed with "two lifesize mannequins in full orthopedic rig . . . [displaying] discreet apertures provided for whatever natural functions were still left to these hybrid creatures" (Ballard, 2001 109).

9. Jean-Luc Nancy points out that "the truth of the play of mirrors must be understood as the truth of the 'with.' In this sense, 'society' *is* 'spectacular'" (2000, 68). Or to be more precise, "There is no society without the spectacle because society is the spectacle of itself" (67).

10. Nabokov seems to anticipate Ballard's post-empathy socioscape in *Lolita* when the two protagonists silently stare "at some smashed, blood-bespattered car with a young woman's shoe in the ditch." As they tire of the spectacle and drive on, Lolita notes, "That was the exact type of moccasin I was trying to describe to that jerk in the store" (1991, 174).

11. In his review of *Super-Cannes* for the *Observer*, Tim Adams writes, "More than any other writer Ballard understands the transformation technology may effect on human desire"—as if the two are somehow separable. It is perhaps also worth noting, given the impact that Michel Houellebecq's novel *Atomized* had on the French public scene, that it also feels drawn to southeastern France as ground zero for a new form of alienation. In contrast to Ballard's executives, however, Houellebecq's characters prefigure a new "asexual and immortal" humanity, "a species which had outgrown individuality, individuation and progress" (2001, 371). The reactionary utopianism of this "vision"—whereby the reader witnesses "a shift which would credibly restore a sense of community, of permanence and of the sacred" (376)—is revealed in all its neo-humanistic glory in the novel's ludicrous epilogue.

12. Heidegger's much earlier critique of tourism is as pithy and prescient as it is sarcastic, for he claims that the modern Rhine, dammed up into the "unreasonable demands" of the hydroelectric power plant, has since been reduced to "an object on call for inspection by a tour group ordered there by the vacation industry" (1977, 16a). Of course, the same could be said for everywhere from Easter Island to the Pyramids of Giza to San Francisco's Fisherman's Wharf.

13. For more on this theme, see Giorgio Agamben's "We Refugees" (1994).

14. Let it be noted that the fictional Eden-Olympia is based on the actual, and equally well-named, Sophia-Antipolis, also located near Cannes.

15. See Jean-Luc Nancy's *Being Singular Plural* (2000, 75).

16. The *Freedom Ship* is a gargantuan ship, described as a "floating city," slated to be launched in 2009. Currently being assembled off the Turkish coast, *Freedom Ship* is selling space to people who can afford to live on this luxurious behemoth, incessantly navigating the world, dwelling everywhere and nowhere.

17. For instance, recently, I, an "Anglo-Australian," bought a Taiwanese toy bunny from a Bangladeshi vendor in Rome—the notable point not being the complex juxtaposition of peoples and places but the fact that we still feel obliged to describe such encounters via cultural citizenship, however displaced.

4. FACING THE INTERFACE

1. The same phenomenon can occur between people who share the same language, as often happens during those moments in a long corridor where you recognize a colleague or acquaintance and yet pretend not to until the last moment. This charade allows you to avoid the ontological shame which accompanies the prolonged, wordless recognition of the other, the sense of being *de trop*.

2. Levinas is largely silent on the differences inherent in an original "After you, sir" and an original "Ladies first," as if gender were irrelevant to the shame-being complex.

3. Bakhtin notes that "every literary work faces outward away from itself, toward the listener-reader, and to a certain extent thus anticipates possible reactions to itself" (2000, 257). Interestingly, we are used to thinking of clocks and watches as having faces, but not books, perhaps because the "meaning" of the latter is assumed to be "deeper." And yet books are vulnerable to being "defaced"—indeed, all objects of value are—which suggests they operate within culture according to the logic of faciality.

4. Without having room to explore the notion further, it is perhaps feasible to see an "Eastern" analogue in the Mandarin desire always to "save face," as a social technology bound to honor and family.

5. Similarly, the black holes of eyes, nostrils, and mouth are described as a "central computer" used to facilitate faciality (Deleuze and Guattari 1999, 177). In yet another parallel, Deleuze and Guattari note that "the prehensile hand implies a *relative* deterritorialization not only of the front paw but also of the locomotor hand. It has a correlate, the use-object or tool: the club is a deterritorialized branch" (172). The face is thus "an *absolute* deterritorialization" of the head-body.

6. This is in complete contrast to the logic behind Levinas's statement that "the whole human body is in this sense more or less face" (1985, 97), since for Levinas the face is "the expressive in the Other," whereas for Deleuze "expression" is merely a psychosomatic symptom of facialization and not an index of an intrinsic, universal humanity.

7. This is certainly one explanation for the compulsion to see faces in all sorts of landscapes, such as "the man in the moon" and "the face on Mars." Such a compulsion carries over to politics, as inscribed in the twentieth-century desire for "socialism with a human face"—as opposed to, say, socialism with great buns and rock-hard abs.

8. Perhaps this screening process awaits the Call, this being the unnamable, untraceable call (a form of ontological telemarketing) that Zizek is so anxious to receive. For Zizek's thoughts on John Woo's *Face/Off*, see *Did Somebody Say Totalitarianism.*

9. Sophisticated facial-scanning software is now poised to overtake the polygraph as the standard lie detector test for government and associated agencies.

The digital procedure involved, however, rests on the same analog assumption that guilt or shame can somehow be read on or through the text of the body, as if breaking the law cannot help but produce signs of transgression within the body language and even the most minute facial expressions of the perpetrator. This is especially significant if we read such soon-to-be standard procedures through the Voight-Kampf test in Ridley Scott's *Blade Runner* (1982), an attempt to distinguish humans from replicants. Apparently, a sense of guilt has replaced memory as the criterion of being human.

10. Screened on *BBC Prime*, Tuesday, March 19, 2002.

11. See Alice's conversation with Humpty Dumpty in *Through the Looking-Glass* (Carroll 1992) for similar sentiments in terms of the face as overrated index of individuality.

12. The medical definition of *chimera* concerns statistically relevant cases of people who in fact carry the DNA of their twin, whose body (and its DNA) was absorbed into the survivor during the gestation process and who thus "lives" on in the other in both an uncanny and genetic sense.

13. "[I]ndividuals are encouraged to identify themselves with their own preferences, to assert them as rights to themselves; all that is expected of them is that they declare their identities and make them available by communication" (Luhmann 1996, 67). As we have seen, Agamben considers the coming community to refuse such self-defeating self-designations.

14. See the work of Japanese roboticist Masahiro Mori (1970), who conducted psychological experiments in order to measure human responses to robots "of varying degrees of anthropomorphism," leading to the AI concept of "uncanny valley." This concept is based on an observeable tipping point whereby humans switch from being charmed by a robot, to being repulsed, once it crosses a convincing mimetic threshold.

15. Dreams and dream-reading play a crucial role in decoding Haruki Murakami's metaphysical cyberpunk novel *Hard-Boiled Wonderland and the End of the World* (1993), in some ways an interesting companion piece to *Neuromancer*. Murakami's protagonist is blinded by a knife in order to read the dreams of the beast skulls which fill the town library. Like Laney, this sightless form of reading is abstract, de-focused, and nodal, since—*contra* Freud—rational or systematic comprehension is not its goal. Perhaps a reading of the space between Murakami's sheep-men, shadows, and unicorns, on the one side, and Gibson's cyborgs, on the other, would help us respond to Philip K. Dick's still-unanswered question: "Do Androids Dream of Electric Sheep?" (1999).

16. The Roman poet-philosopher Lucretius (c. 100–55 B.C.) believed that images are material "films" which constantly fly off the surface of objects and strike the eye, causing vision. In an uncanny anticipation of both Baudrillard's "precession of the simulacra" and Deleuze's "all the world turns to film" (1995, 76), Lucretius discusses the human reflex to see in the clouds "giant faces [which] appear to be sailing by" (1994, 98). These fluid images belong to a

different category of film than those which emanate from objects, since they are "spontaneously generated"—not unlike the Idoru herself. (See also p. 115 for a discussion of dream images which seem to anticipate the process of animation as well as the optical trickery of 24 frames per second.)

17. From a certain perspective, the Idoru's ontological status could be considered a technosecular form of Mark 5:9 in the Bible: "And he asked him, What is thy name? And he answered, saying, My name is Legion: for we are many" (King James Version).

18. In a different work, "Repetition and Stoppage: Guy Debord's Technique of Montage," Agamben discusses the disconcerting cinematic novelty of Ingmar Bergman's film *Monika*, in which the main character (Harriet Andersson) suddenly stares directly at the camera. Quoting Bergman directly, he notes, "Here, and for the first time in the history of cinema, a direct, *shameless* contact is established with the viewer" (1996, 71, my emphasis). Agamben notes, however, that pornography and advertising have since normalized this procedure.

19. Differences could be accorded between stage fright, a shame associated with being exposed to people, and "studio fright," in which the shame is prompted by exposure to machines. For instance, it is said that Josephine Baker, who never suffered from stage fright, would freeze up in the studio if the technicians became too engrossed in their work. A similar tension lies behind the function of the "Humanize" button, found on old musical sequencers, which was included in order to make "clinical," metronomically perfect beats more erratic (despite being based on an "algorhythm").

20. Contrast with Montaigne's fable of the beggar who wears only a shirt in the depths of winter. When asked by a fur-covered gentleman how he can smile in such conditions, the beggar replies, "But you, sir, have your face exposed. Well, I am all face" (1970, 121). Indeed, the poverty-stricken are but one example of *les indévisageables* (those we cannot look in the face).

21. Such intense interpellation—between face, name, subjectivity, and identity—prompts Foucault to state, "I am no doubt not the only one who writes in order to have no face" (1972, 17).

22. http://www.digimask.com. See also Mark Hansen's article "Affect as Medium, or The 'Digital-Facial-Image.'"

23. See also the MIT language and gesture group: *http://www.media.mit.edu/groups/gn/*. Thus far Digimask's range seems quite limited, since its publicity literature cites "smiley, nervous, angry, tired" and something called "hero" as the only available options of expression.

24. Of course, the doctoring of photographs has been around since the dawn of this particular medium. However, the internet, along with image manipulation software, has created a cult for "celebrity pornography," in which "legitimate" star's features are inscribed onto porn stars' faces. See also Clemens and Pettman, "From the 7–11 to September 11: Popular Propaganda and the Internet's War on Terrorism" (2004).

25. In an era where this term is used to sell enormous quantities of old Enya CDs, one genuinely wonders whether the concept of "passion" has been drained of any and all rhetorical power.

26. "Inasmuch as it is nothing but pure communicability, every human face, even the most noble and beautiful, is always suspended on the edge of an abyss" (Agamben 2000, 96).

27. To avoid potential misreadings of my tone here: I am not criticizing the notion of the couple itself but one of its more common incarnations, something I referred to in an earlier chapter as the "bunker mentality." The question is whether it is in fact possible to maintain a "healthy" exposure to other people—outside the couple—without resorting to the rather domesticated alternatives of "swinging" or an "open relationship." Both are notoriously difficult to maintain, even when they are considered desirable modes of being rather than simply "lifestyle choices." For straddling the other side of this seesaw of denial, across from the bunker couple, is the egocratic single person who refuses to "get hitched" (as if it were really viable to live as a singular versus the plural). In other words, it seems, on the one hand, important to resist the traditional, institutionalized injunction to settle down into an entropic, obedient, and eventually resentful couple. Yet, on the other hand, it seems equally important to resist the constant command of the (meat) market to be a restless, thoughtless, and promiscuous consumer of other people, treating them like the latest gym equipment ordered over the telephone during a drunken and fleeting moment of imagined repentance.

28. "Being-social is Being that is by appearing in the face of itself, faced with itself: it is *co-appearing*" (Nancy 2000, 59).

5. "HOW WAS IT FOR ME?" NOT-SEEING THE NON-SPACES OF PORNOGRAPHY

1. Film critic Amy Taubin defends this scene on the grounds that Kubrick died before finishing the postproduction, suggesting that he would never have allowed it to be released in its current form. Indeed, the director was famous for his meticulous and marathon attention to this part of the filmmaking process, and one can only speculate whether this scene could have been "salvaged" through better editing (not to mention through undoing the unauthorized "censorship" inflicted on American prints).

2. For more on Maffesoli and the orgiastic metaphor, see my book *After the Orgy: Toward a Politics of Exhaustion* (2002).

3. Thomas Edison himself was a pioneer pornographer, producing *The Trapeze Disrobing Act* and other "smoking concert" films (in Allen, 267–68). See Linda Williams's book *Hard Core: Power, Pleasure, and the "Frenzy of the Visible"* (1990) as well as Lynn Hunt's edited volume *The Invention of Pornography: Obscenity and the Origins of Pornography, 1500–1800* (1996), for more thorough treatments of the emergent conditions of porn media histories.

4. See Walter Kendrick's *Secret Museum* (1996) for a detailed history of the changing criteria used to distinguish one from the other, especially his conclusion that "the kind of 'hard-core' pornography we now would place at the bottom of the social scale belonged at the top a century ago. Its quality was no higher than what we are familiar with today, but its circulation was confined to that class of 'safe' readers who were granted easy admission to the age's other Secret Museums. It therefore figured hardly at all in public controversies; there was, literally, no harm in it" (Allen 1991, 302).

5. The attempt to distinguish between "art" and "pornography" often includes a notion of the "invisible remainder," implying that "pornography" has little symbolic surplus over and above the image. "Good" porn has an invisible remainder, the argument goes, and thus actually transcends itself in order to become "erotica" or "art." Ergo, there is no such thing as good pornography.

6. This would classify Blake's output as "stag films," according the distinction proposed by Robert C. Allen: "Unlike the hard-core feature film, whose pleasures are predicated on providing narrative and visual satisfaction, the stag film's pleasurability resides in its ability to arouse but not to satisfy the viewer" (1991, 270).

7. "*Cahiers du Cinéma* says the porn movie is the true future of film" (Ballard, 2001, 285).

8. Heidegger's comments in "Letter on 'Humanism'" (1998) on the crucial role of boredom could apply to an Andrew Blake movie.

9. On the resonance between these two crystallizations of capital, Agamben writes, "The fact that the [porn] actors look into the camera means that they *show that they are simulating;* nevertheless, they paradoxically appear more real precisely to the extent to which they inhabit this falsification. The same procedure is used today in advertising: the image appears more convincing if it shows openly its own artifice" (2000, 94).

10. To borrow a Lacanian pun from David Odell, pornography is all about the $.

11. See Steven Soderbergh's *Sex, Lies and Videotape* (1989), for an exploration of this kind of "meta-alienation," as well as the study by Brian McNair (1996) on the erotics of mediation in postmodernity.

12. For a more sophisticated treatment of the question of cybersex, and mediated communication in general, see Pierre-Paul Renders's film *Thomas est Amoureux*, an innovative and strangely intimate reverse portrait of a sociopath who can make love only via Visiophone.

13. Interestingly, the first browser designed for the World Wide Web was called Mosaic.

14. Commenting on the shifts in the lover's discourse during the eighteenth and nineteenth centuries, Niklas Luhmann notes: "The search for a new unifying formula for love, sexuality and marriage and the location of this in the idea of personal self-realization ultimately had consequences for the area of pornog-

raphy and the obscene. What could be rejected in the name of these concepts now had to be restricted and accordingly is weighed down with complications which could then once again stimulate a desire to attempt to balance on the border between them and love. . . . [This last is because] it was not yet possible to draw a line separating sensuality and the soul if the unity of both is required for love and marriage. The obscene discounted itself owing to the fact that it occluded an interest in the person, or more precisely, owing to the exchangeability of the persons involved. . . . The orientation towards the individuality of the partner . . . [thus] came up against the limits of what could be preprogrammed by cultural codification" (1998, 119–20).

15. The first definition of "machine" in the *Oxford English Dictionary* is as "a structure of any kind, material or immaterial; a fabric, an erection." As early as 1749 John Cleland's text simply titled *Fanny Hill: Memoirs of a Woman of Pleasure* contains the phrase "Coming out with that formidable machine of his, he lets the fury loose."

6. A SELF OF ONE'S OWN?

1. Impersonators make a living from the unconscious recognition that every person is, to a great extent, im-personal.

2. Consider Clifford's words from *Lady Chatterley's Lover:* "The more I live, the more I realise what strange creatures human beings are. Some of them might just as well have a hundred legs, like a centipede, or six, like a lobster. The human consistency and dignity one has been led to expect from one's fellow-men seem actually non-existent. One doubts if they exist to any startling degree even in oneself" (Lawrence 1994, 266).

3. That is to say, the media fusions of Ben Affleck and Jennifer Lopez, as well as Tom Cruise and Katie Holmes. Or as Virginia Woolf puts it in *The Waves:* "To be contracted by another person into a single being—how strange" (1978, 89).

4. "[V]iewpoints toward a world supposedly the same are as different as the most remote worlds. This is why friendship never establishes anything but false communications, based on misunderstandings, and frames only false windows. This is why love, more lucid, makes it a principle to renounce all communication. Our only windows, our only doors are entirely spiritual; there is no intersubjectivity except an artistic one" (Deleuze 2000, 42). (This is Deleuze's take on Lacan's statement "Il n'y a pas de rapport sexuel," which Deleuze has significantly truncated to "Il y a *uniquement* de rapport sexuel.")

5. Anticipating the link a little, Wall states: "There is nothing mysterious, magical, or ineffable about the Whatever. It is as common as can be. It is the most common" (1999, 123).

6. We could refer here to the myth of Theseus's ship, which was repaired so extensively over time that eventually no original piece was left. The metaphysi-

cal question then becomes, Is this still Theseus's ship?—a question we could apply to our own bodies *qua* identity every seven years or so, given the time scale of cellular regeneration.

7. Later in the same passage, Deleuze writes that "it is true that our loves repeat our feelings for the mother, but the latter already repeats other loves, which we have not ourselves experienced. . . . At its limit, the experience of love is that of all humanity, which is traversed by the current of a transcendent heredity" (2000, 72).

8. "The entire Search sets three kinds of machines to work in the production of the book: machines of partial objects (impulses), machines of resonances (Eros), machines of forced movement (Thanatos)" (Deleuze 2000, 160).

9. Regarding this feedback loop, Nancy writes that "meaning does not consist in the transmission from a speaker to a receiver. . . . As far as meaning is concerned, what I say is not simply 'said,' for meaning must return to me resaid in order to be said" (2000, 86).

10. "Usually man does not show his body, and, when he does, it is either nervously or with an intention to fascinate. . . . Shame and immodesty, then, take their place in a dialectic of the self and other which is that of master and slave" (Irigaray 2000, 20). We are back in the Austro-literary realm of underwear and uniforms here.

11. "Let us call 'subject' every finite state of a generic procedure" (Badiou in Cadava et al. 1991, 26), a quote from Cadava's collection, *Who Comes after the Subject*, ironically enough, a veritable who's who of early-1990s Continental philosophy.

12. "[E]xistence is standing out and perduring the openness of the there: Ek-sistence" (Heidegger 1996, 125). Or, as Lewis Carroll's Duchess puts it: "Never imagine yourself not to be otherwise than what it might appear to others that what you were or might have been was not otherwise than what you had been would have appeared to them to be otherwise" (1992, 72).

13. See also *Being Singular Plural* (Nancy 2000, 30–31, 93), where *Mitsein* is criticized for being subordinate to *Dasein*.

14. Indeed, Heidegger's notion of concern intersects with Levinas's ethics of alterity when he confirms that *Dasein* "is" essentially for the sake of others (1996, 116), and a little later, "Everyone is the other, and no one is himself" (120). Contemporary theorist Cameron Bailey extends this idea through a Bantu saying: "Umntu ngumntu ngabantu—a human being is a person through (other) people" (339).

15. Perhaps I have left it long enough before mentioning Heidegger's infamous sympathy to Nazism, given that it must always be taken into account when discussing both the text and context of his ideas. One of the standard strategies of his supporters is to say the same thing about Heidegger's philosophical edifice that he himself says about modern technology: "There where the danger lies, so grows the saving power." However, a conscious decision this

fraught and complex should not be dismissed with a maxim which becomes more glib with each repetition. This is why I refer the reader to the excellent film *The Ister* (2004), which does a commendable job not only of presenting Heidegger's ideas in an accessible manner but also in contextualizing ("struggling with") the legacy of his "unpardonable" "criminal" "involvement" (Nancy 2000, 26) and silence during the 1930s and '40s.

16. Prefigured by Nietzsche's insightful question: "How is explanation to be at all possible when we first turn everything into a picture—our picture!" (2001, 113).

17. Note the difference in emphasis in Deleuze's thinking: "By singularity, we mean not only something that opposes the universal, but also some element that can be extended close to another, so as to obtain a connection; it is a singularity in the mathematical sense" (1991, 94).

18. This in contrast to English pop singer Sam Brown's lyric "I am all I know, so I must know me well."

19. See Bruce Mazlish's *The Fourth Discontinuity* (1993), which traces the four great ego bruises to the human race, from Copernicus-Galileo, through Darwin and Freud, to the technologies of artificial intelligence.

20. "Love is thus not, as the dialectic of desire suggests, the affirmation of the self in the negation of the loved object; it is, instead, the passion and exposition of facticity itself and of the irreducible impropriety of beings. *In love, the lover and the beloved come to light in their concealment, in an eternal facticity beyond Being*" (Agamben 1999b, 204).

21. For a particularly sharp exegesis of this formal emptiness, see Paola Marrati's "Against the Doxa: Politics of Immanence and Becoming-Minoritarian," in which she writes, "The problem of the majoritarian model is that it produces norms of identification which impose themselves on the whole world and yet represent no one" (2001, 210).

22. "[Soldiers submit] to the preemptive needs of the Manœuvre—a Soldier's Faith at last must rest in the Impurity of his own desires. What can Hansel possibly wish for, that Heinz in front of him, and Dieter behind, and a couple of Fritzes on either side, have not already desir'd,—multiplied by all the ranks and files, stretching away across the Plain? The same blonde from down the Street, the same Pot of beer, the same sack of Gold deliver'd by some Elf, for doing nothing. Who is unique? Who is not own'd by someone? What do any of their desires matter, if they can be of no use to the Manœuvre, where all is timed from a single Pulse, each understanding no more than he must" (Pynchon 1998, 551).

23. Agamben, in a typically "meta" gesture, uses the concept of an example as an example itself for the figure of whateverbeing: The example "is one singularity among others, which, however, stands for each of them and serves for all. . . . Neither particular nor universal, the example is a singular object that presents itself as such" (1993a, 10).

24. For a closer reading of this remarkable film, see Simon Critchley's article "Calm: On Terence Malick's *Thin Red Line*."

25. For the ultimate statement on *Gravity's Rainbow*—at least in terms of the textual troping of ontology—see Leo Bersani's "Pynchon, Paranoia, and Literature," in which he states, "Through Slothrop we mourn the loss of personal presence, of a myth of personality that may, after all, be the only way in which our civilization has taught us to think about ourselves (to think our selves), a loss that must nonetheless be sustained if we are also to disappear as targets" (1990, 193). Further, he argues that "Pynchon's novel is a dazzling argument for shared or collective being—or, more precisely, for *the originally replicative nature of being*. Singularity is inconceivable; the original of a personality has to be counted among its simulations. Being in Pynchon is therefore not a question of substance, but rather of distribution and collection" (194). Unfortunately, I do not have the space here to elaborate on Bersani's tagging of love as an "extra-paranoid myth" (190), but I hope to do so in another context.

26. This is a point which should be kept in mind given the street use of the term "whatever" in alternative American youth culture. See particularly Kurt Cobain's "Whatever nevermind" in Nirvana's instant classic track "Smells Like Teen Spirit" (1991), and Liam Lynch's even more ironic "The United States of Whatever" (2002).

27. Heidegger himself links all these elements in this comment from "Letter on 'Humanism'": "It could even be that nature, in the face it turns toward the human being's technical mastery, is simply concealing its essence" (1998, 247).

28. "As soon as there is love," writes Nancy, "the slightest act of love, the slightest spark, there is this ontological fissure that cuts across and that disconnects the elements of the subject-proper. . . . One hour of love is enough, one kiss alone" (1991, 96).

29. See Agamben's reading of this moment in Antelme, in the "Shame, or On the Subject" chapter of *Remnants of Auschwitz* (1999a).

7. MIND THE GAP

1. Presumably a *womanufactured* world would be less alienating, by Irigaray's account.

2. This is something which could have been more easily avoided had she used Bakhtin rather than Hegel to make her point.

3. N. Kathryn Hayles dates the posthuman quite differently, when she writes: "The important intervention comes not when you try to determine which is the man, the woman, or the machine. Rather, the important intervention comes much earlier, when the [Turing] test puts you into a cybernetic circuit that splices your will, desire, and perception into a distributed cognitive system in which represented bodies are joined with enacted bodies through mutating and flexible machine interfaces. As you gaze at the flickering signifiers

scrolling down the computer screens, no matter what identifications you assign to the embodied entities that you cannot see, you have already become posthuman" (1999, xiv). However, this more historical narrative, depicting the posthuman as displacing the liberal subject, is developed on the level of cultural discursive construction, rather than ontology. That is to say, Hayles focuses more on the moment we *realized* we were *toujours déja* posthuman, a moment delayed by many millennia.

4. Moreover, the same can be said of text and hypertext—and perhaps, on an even grander scale, of physics and metaphysics.

5. Granted, the (post)apocalyptic scenario forces us to think of a return to previous technological eras. Depending on the scale of devastation, we will be thrown back to the Stone, Bronze, or Commodore 64 Age. But what *Mad Max 2* (1981), *A Canticle for Leibowitz* (1960), *Riddley Walker* (1980), and countless other similar tales show is that this can never be figured as an "innocent" return, that the memory and logic of the "more advanced past" contaminates this new-old present, like radiation.

6. "We are operating only with things that do not exist—with lines, surfaces, bodies, atoms, divisible times, divisible spaces. . . . [I]n truth a continuum faces us, from which we isolate a few pieces, just as we always perceive a movement only as isolated points, i.e. do not really see, but infer. . . . An intellect that saw cause and effect as continuum, not, as we do, as arbitrary division and dismemberment—that saw the stream of the event—would reject the concept of cause and effect and deny all determinedness" (surprisingly, perhaps, not Deleuze, but Nietzsche [2001], 113).

7. It is still open to debate whether heterosexually identified people possess gaydar, since many would maintain that it is a technology used exclusively for and by homosexuals. Personally, I believe everyone has gaydar, albeit attuned to different degrees of sophistication and activating different responses. Indeed, it was *Sex in the City*–type women who popularized the term in the late 1990s while using their faculties to sort newly encountered males into either "friend" or "potential lover" categories.

8. See *http://www.bioneers.org*.

9. We can only speculate on the spectacle that would result from allowing the athletes to take whatever substances they desire, since it would produce both angels and monsters, humans and cyborgs—that is to say, genuine experiments in "interior design."

10. See Waugh's novel *Decline and Fall* (1928).

11. In his paper entitled "The Turning," Heidegger maintains, "Technology will not be struck down; and it most certainly will not be destroyed" (1977c, 38); before insisting, "All that is merely technological never arrives at the essence of technology. It cannot even once recognize its outer precincts" (48).

12. For a discussion of the "objectum sexuality," the erotic relationship between humans and objects, see J. Clemens and D. Pettman, "Relations with Concrete Others," in *Avoiding the Subject* (2004). This theme is also treated in

Marco Ferreri's film *I Love You* (1986), in which a man falls hopelessly in love with a keychain.

13. See also Heidegger's thoroughly techtonic comments concerning mountain ranges (1977a, 19).

14. There is a latent Marxist narrative in Heidegger's take on technology, and not only when he attacks the desire of our species to exploit "the maximum yield at the minimum expense" (1977a, 15). We find another parallel in the form of anticipation these two German heavyweights accord to history, whereby Capital or Enframing reaches a critical mass (although Heidegger waits more for revelation than revolution), when the latter one says that "the frenziedness of technology may entrench itself everywhere to such an extent that someday, throughout everything technological, the essence of technology may come to presence in the coming-to-pass of truth" (35). Moreover, Heidegger also writes against that technological formation dubbed "the Spectacle" by neo-Marxists like Guy Debord, claiming that our "hearing and seeing are perishing through radio and film under the rule of technology" (48)—a claim which unfortunately recuperates Heidegger as something of a Luddite himself, albeit an exceedingly complex instance of one.

15. As Werner Heisenberg cried prophetically—living up to his own principle of uncertainty—"But I don't even know what a matrix is!" (in Pesic 2002, 136).

16. See Andy Clark's *Natural-Born Cyborgs* (2003), which presents intelligence as a "distributed system"—that is, belonging to neither humans nor machines exclusively, but in between. A companion piece to this is the work of Timothy Luke (1999), who unmasks the closet humanism of commentators such as Nicholas Negroponte (1995), who discuss "being digital" at the expense of "digital beings." Finally, it is clear from television interviews that the Russian cosmonauts on the *Mir* space station believed their "mother ship" to be "a living thing" with whims and intentions of its own (see the BBC Horizon documentary *Mir Mortals* [1998]).

17. For example, even certain caterpillars, which crawl unheedfully through the dirt, are called "inchworms" by apprentices of the symbolic order—the same who mistakenly believe not only that it is "measuring the marigolds" but that "you and your arithmetic, you'll probably go far."

18. From this perspective the troubadours embody an historically active line of transmission.

19. Significantly or not, the most devastating computer virus of 2000 was the I Love You virus.

20. I use the word *incarnation* advisedly, since Björk's droids have no material link to the fleshy etymology of *in-carnation.*

8. ASYMPTOTIC ENCOUNTERS: LOVE FREED FROM ITSELF

1. Michel de Certeau begins his essay "Walking in the City" (1988) from this now-vanished location, creating a proleptic and unsettling effect in the

mind of the post–September 11 reader. Moreover, his decision to include a pictorial reproduction of an advertisement for the World Trade Center observation deck, featuring the slogan "It's hard to be down when you're up," now functions in hindsight as a darkly ironic warning to the doomed tenants.

2. These patterns can be quite beautiful, as represented by the charts commissioned by those in charge of "event security" or designing spaces for large crowds.

3. This is emphatically *not* to romanticize the latter *against* the former, à la certain simplifications of de Certeau's notion of "tactics."

4. For a longer discussion of Wong Kar-wai, see Clemens and Pettman, "The Floating Life of Fallen Angels," in *Avoiding the Subject* (2004).

5. Perhaps Wong is thus partially to blame for the 2002 Lacoste television advertisement which shamelessly exploits the latent Orientalism offered by this particular film.

6. When I say "Wong Kar-wai," I am in fact discussing an amalgam or avatar which, besides Wong, includes his regular actors, Tony Leung and Maggie Cheung, as well as his cinematographer, Christopher Doyle, who more elegantly than most "writes in light."

7. In Nancy's terms, "not, perhaps, an experience that we have, *but an experience that makes us be*" (1991, 26).

8. As if speaking directly of Wong Kar-wai, Levinas states, "I do not see what angels could give one another or how they could help one another" (1985, 97). Lacan, on the other hand, is not quite so dismissive, producing his own possibilities through his usual wordplay "*étrange, étre-ange*" (strange, angel-being) (1999, 8).

9. My thanks to Megan Quinlan for her perceptive reading of this scene, as well as for her insights regarding the relationship between consumer and romantic choice in Wong Kar-wai's films (and indeed, society in general).

10. Murakami translated these authors into Japanese, and their general style have inevitably left an impression on his own work.

11. We should, however, guard against repolarizing the world into West versus East, either figured as a soliloquy or a dialogue, especially after Ella Shohat and Robert Stam's deployment of "polycentric multiculturalism." Indeed, connecting back to one of our paradigmatic figures, these same writers note that the "Brazilian essayist, poet, novelist, anthropologist, musicologist, Mario de Andrade" gave his 1928 novel *Macunaima* the subtitle *The Hero without Any Character*. Where de Andrade's protagonist differs from Musil's is perhaps in an excess of "culture" (i.e., hybridity), rather than the latter's almost existential lack of it.

12. In a quote which collapses the distinction between organics and artifice, Agamben notes that the French term *faitis*, "like its German counterpart, *feit*, simply means 'beautiful, pretty.' In particular, it is used in conformity with its etymological origin to designate that which, in a human body, seems made by

design, fashioned with skill, made-for, and which thereby attracts desire and love" (1999b, 196).

13. Editorial from http://www.amazon.fr/exec/obidos/ASIN/1576870936/qid=1150209399/sr=8-2/ref=sr_8_xs_ap_i2_xgl14/402-8669894-4116905.

14. The improbable quest of William Gibson's "footage-heads," who trawl the internet for traces of an enigmatic "garage Kubrick" in *Pattern Recognition* (2003), makes a lot more sense when read in light of this quote by Agamben.

15. The most lamentable instance of these aspirational style guides is England's *Wallpaper** magazine, for its complete, vulgar, and sincere surrender to the aestheticization of everything. Rather than simply recoiling in horror at the world-historical nihilism underwriting this venture, we would do well to embrace our most optimistic instincts and see it as the ignorant herald of this system's complete exhaustion—a grotesque angel flying backward through the catastrophe.

16. Here I'm thinking particularly of Nietzsche's definition of *nihilism*, where it is a synonym for *pathos* or *Christianity*.

17. "[T]hough no one seems to call the advertising world 'Madison Avenue'
anymore. Have they moved?
I need an update on this."

From David Berman's poem, "Self-Portrait at 28" (1999, 60)

18. In 1967, before this generation of SMS addicts was born, the Department of Information at IBM issued a list of roles that the computer could play in daily life, including (1) as diagnostician, doctor, and druggist; (2) as Cupid, searching the files for ideal matches; (3) as marriage monitor and arbiter; and (4) as immigration counselor (McLuhan and Fiore 1997, 94).

19. Interestingly, *pixilated* (with a second *i*) used to mean "crazy" or "drunk"—as from "off with the pixies."

20. Mary Douglas's influential definition of *danger* is that which cannot be confidently categorized.

21. Unfortunately, the movie then backpedals somewhat from this wonderful acknowledgment of being purely "whatever," succumbing to the Hollywood injunction for reestablishing family roots. The film, however, then pedals forward once again at the dénouement, when Gonzo rejects his newly found extraterrestrial family for his own motley Muppet community. Complicating things further, Kermit the Frog undoes the progress made with this one word "whatever" by redefining it as "distinct," in direct contrast to the way we have been employing the term. (Thanks to Patrick van Schaik for this example.)

22. "[T]he whole of nature is one individual, whose parts, that is, all bodies, vary in infinite ways, without any change of the whole individual" (Spinoza in Pesic 2002, 126).

23. See the writings of Hermann Weyl for his discussion of Mike and Ike, the "quantum twins," whom we can never, even in principle, distinguish from

each other. As Pesic notes in his commentary, Weyl was moved to note that "one cannot demand an alibi of an electron!" (2002, 98)—a comment with seemingly infinite juridical and ontological implications.

24. It is not only physics that is redefining current techniques of understanding ourselves and our world, but also the medical and biological sciences. Following Alice through the looking glass, researchers have identified the existence of "mirror neurons" in the brain, which "fire" even when they are not connected to an actual object or stimulus. For example, if I watch someone getting an injection, my mirror neurons will fire as if I "myself" were receiving the injection. The theory is that this kind of identification (or empathy) is an evolutionary mechanism to allow us to mimic certain behaviors and tasks. However, it also poses interesting metaphysical questions about the I who is mirroring the other person. If my brain "does" something when my body is passive—by proxy, as it were—then on what level am I an autonomous, sovereign creature, disconnected from my neighbor? Well, the answer is, of course, less than I may think.

25. Judith Butler states, "This is not to say that there is no foundation, but rather, that wherever there is one, there will also be a foundering, a contestation" (51).

26. Paul Valéry: "My fate is more me than myself. A person is only made up of answers to a number of impersonal incidents" (in Agacinski 1991, 21).

27. Take, for example, the phrase "I am enjoying myself." What, exactly, is the object of enjoyment here? Is it identical with the subject? Is "enjoying myself" a simple reflexive verbal expression, as is "washing myself"? And if so, shouldn't the host of a party be offended if a guest claims to be enjoying *him- or herself,* since the guest could be doing that anywhere? Shouldn't, rather, he or she be enjoying the party?

28. Of course, the aporia here, through which all prescriptive arguments fall, is the rhetorical category of the "we"—specifically, those who are included and excluded.

29. Dan Ross puts a positive spin on narcissism, claiming that "it is not a deafness or blindness to others, but the very condition of being-with-others. Primordial self-love is the capacity without which the extension of love to others is impossible." Such a disagreement comes down to definition, for I deploy the term as the psychological condition which blocks the possibility of appreciating the important statement that Ross immediately goes on to make: "*I* and *we* are not substances, not identities, but individuations in the course of becoming, without stability if nonetheless capable of metastability. Achieving metastable rather than stable equilibrium, *I* and *we* are always unfinished, never quite identical to themselves, always more or less *out-of-phase* with themselves, conserving themselves *in the form of* permanent individuation. They are therefore capable of becoming, of *movement*. . . . [T]his process of permanent individuation must not be reduced to that entropic becoming of the universe described by physical

laws. Human individuation is, on the contrary, *negentropic*, a matter of *life*" (2006, 2).

CONCLUSION. OF MICE AND MULTITUDES

1. Admittedly, Robins usefully reminds us of "Winnicot's notion of potential space: the 'third space of human living,' neither inside the individual nor outside in the world of shared reality, the space of creative playing and cultural experience" (2000, 85). However, this is not deployed to full effect in Robins's own polemic.

2. It is impossible to explain the hegemony of Christianity, for instance, without reference to the interpenetration of technology (the Cross, revelation, communication), love (God's word, affection for thy neighbor), and community (the Church, the Saved, the flock). This same rule applies to any other grand narrative, for that matter.

3. A century in which, to quote Peter Parker's uncle in the movie *Spider-Man* (2002), "even computers need analysts."

4. A subcriticism concerning conflation could be summarized in the sentence "I would certainly prefer talking to a human than a computer on a train." My point, however, is that we cannot discount the role of the train—nor indeed of the talking.

5. Notice that we *take* pity, rather than give it.

6. The strength and popularity of Hardt and Negri's *Empire* (2000) was due to an insightful understanding of the relationship between ontology, economics, and materialist political forms of resistance. Unfortunately, their project shoots itself in the foot when it assumes capital will somehow run its course or, failing that, advertising executives and graphic designers will suddenly wake up to the machinations of ideology and trade in their champagne glasses for Molotov cocktails.

7. In keeping with the self-leasing metaphor, we have several options to take. One is to trash and thrash our bodies like a rented car (as rock stars tend to do). Another is to treat it with a certain amount of critical "care," as Foucault advocated later in life—at least in theory.

8. Thus, when I say "we," I use it more in the spirit of the French informal pronoun *on*, which can mean both "we" and "they."

9. Heidegger considers love and hate as special cases—passions, which form a fundamental circuit within *Dasein*, as opposed to fleeting, sparking affects.

10. Nietzsche is thinking specifically of the "erroneous assertion" that the human character is unalterable, for "if we were to imagine a human being eighty thousand years old, we would have in him an absolutely alterable character: so that an abundance of different individuals would gradually develop out of him" (1995, 49).

11. Burroughs is quoted in Shaviro, "Returning to the Scene of the Perfect Crime, Or, How I Learned to Stop Worrying and Love the Virtual." This

unpublished paper also points out the interesting position of the Catholic Church on cloning—specifically, that "this duplication of body structure does not necessarily imply a perfectly identical person, understood in his ontological and psychological reality" (2004). In stark contrast, Deleuzians actively seek a different point, "not the point where one no longer says I, but the point where it is no longer of any importance whether one says I" (Deleuze and Guattari 1999, 3).

WORKS CITED

Abbas, Ackbar. 1997. *Hong Kong: Culture and the Aesthetics of Disappearance*. Minneapolis: University of Minnesota Press.

Adams, Carrie Olivia. 2006. "Vermilion." In *A Useless Window*. Boston and Chicago: Black Ocean.

Adams, Tim. 2000. "Review of the Year." In *The Observer*. December 31, R4.

Aeschylus. 1961. *Prometheus Bound and Other Plays*, translated by Philip Vellacott. Harmondsworth and New York: Penguin Classics.

Agacinski, Sylvia. 1991. "Another Experience of the Question, or Experiencing the Question Other-Wise." In *Who Comes after the Subject*, translated by M. Syrotinski and C. Laennec, edited by E. Cadava et al. New York: Routledge.

Agamben, Giorgio. 1993a. *The Coming Community*, translated by M. Hardt. Minneapolis: University of Minnesota Press.

———. 1993b. *Stanzas: Word and Phantasm in Western Culture*, translated by R. L. Martinez. Minneapolis: University of Minnesota Press.

———. 1994. "We Refugees." Online. Internet. http://www.europeangraduate school.de/faculty/agamben/agamben-we-refugees.html.

———. 1996. "Repetition and Stoppage: Guy Debord's Technique of Montage." In *Documenta Catalogue*. Ostfildern-Ruit: Cantz Verlag.

———. 1998. *Homo Sacer: Sovereign Power and Bare Life*, translated by D. Heller-Roazen. Stanford, Calif.: Stanford University Press.

———. 1999a. "Shame, or On the Subject." In *Remnants of Auschwitz: The Witness and the Archive*, translated by D. Heller-Roazen. New York: Zone Books.

———. 1999b. "The Passion of Facticity." In *Potentialities: Collected Essays in Philosophy*, translated by D. Heller-Roazen. Stanford, Calif.: Stanford University Press.

———. 2000. *Means without End: Notes on Politics*, translated by V. Binetti and C. Casarino. Minneapolis: University of Minnesota Press.

———. 2004. *The Open: Of Man and Animal*, translated by Kevin Attell. Stanford: Stanford University Press. .

Allen, Robert C. 1991. *Horrible Prettiness: Burlesque and American Culture*. Chapel Hill: University of North Carolina Press.

Anderson, Benedict. 1991. *Imagined Communities: Reflections on the Origin and Spread of Nationalism.* London and New York: Verso.

Armstrong, K. 2000. *The Battle for God.* London: HarperCollins.

Atwood, Margaret. 1969. *The Edible Woman.* Toronto and Montreal: McClelland and Stewart.

Augé, Marc. 1995. *Non-places: Introduction to an Anthropology of Supermodernity.* London and New York: Verso.

Auster, Paul. 1992. *The New York Trilogy.* London: Faber and Faber.

Ayerza, Josefina. 2001. "To Resume Again . . ." In *Lacanian Ink.* Spring. No. 18: 3–4.

Badiou, Alain. 1991. "On a Finally Objectless Subject," translated by Bruce Fink. In Cadava, E., P. Connor, et al., eds. *Who Comes after the Subject?* New York: Routledge.

Bailey, Cameron. 2001. "Virtual Skin: Articulating Race in Cyberspace." In *Reading Digital Culture.* David Trend, ed. London: Blackwells.

Bakhtin, Mikhail M. 1984. *Problems of Dostoevsky's Poetics,* translated by Caryl Emerson. Minneapolis: University of Minnesota Press.

———. 2000. "Forms of Time and of the Chronotope in the Novel: Notes toward a Historical Poetics." In *The Dialogic Imagination: Four Essays.* Edited and translated by M. Holquist. Austin: University of Texas Press.

Ballard, J. G. 1973. *Crash.* New York: Farrar, Straus & Giroux.

———. 2001. *Super-Cannes.* London: Flamingo.

Barthes, Roland. 1990. *A Lover's Discourse: Fragments,* translated by R. Howard. London: Penguin.

Barzun, Jacques. 2004. "Introductory Remarks to a Program of Works Produced at the Columbia-Princeton Electronic Music Center." In Cox, Christoph and Daniel Warner (eds.) *Audio Culture: Readings in Modern Music.* New York and London: Continuum.

Bataille, George. 1985. "The Solar Anus," translated by Allan Stoekl, et. al. In *Visions of Excess: Selected Writings, 1927–1939.* Minneapolis: University of Minnesota Press.

———. 1986. *Erotism: Death and Sensuality,* translated by M. Dalwood. San Francisco: City Lights Books.

Baudrillard, Jean. 1988. "On Seduction." In *Jean Baudrillard: Selected Writings,* edited by Mark Poster, translated by J. Mourrain. Cambridge, England: Polity Press.

———. 1990. *Fatal Strategies,* translated by J. Fleming. New York: Semiotext(e).

———. 1994. *The Illusion of the End,* translated by Chris Turner. Oxford: Polity Press.

Bauman, Zygmunt. 1996. "Morality in the Age of Contingency." In *Detraditionalization: Critical Reflections on Authority and Identity,* edited by P. Heelas, S. Lash, and P. Morris. London: Blackwells.

———. 2003. *Liquid Love: On the Frailty of Human Bonds.* Cambridge, England: Polity Press.

Beck, Ulrich, and E. Beck-Gernsheim. 1996. "Individualization and 'Precarious Freedoms': Perspectives and Controversies of a Subject-Orientated Sociology." In *Detraditionalization*, edited by P. Heelas et al. London: Blackwells.

Becker, Dana. 1997. *Through the Looking Glass: Woman and Borderline Personality Disorder.* Boulder: Westview Press.

Benjamin, Walter. 1992. "Theses on the Philosophy of History." In *Illuminations*, edited by H. Arendt, translated by H. Zohn. London: Fontana.

Benvie, Robert. 2001. "The Porn I Like." *Timothy McSweeney's* (May 8), http://www.mcsweeneys.net/2001/05/08porn.html.

Bergson, Henri-Louis. 1988. *Matter and Memory*, translated by N. M. Paul and W. S. Palmer. New York: Zone Books.

Berman, David. 1999. "Self-Portrait at 28." In *Actual Air: Poems.* New York: Open City.

Bernasconi, R. 1993. "On Deconstructing Nostalgia for Community within the West: The Debate between Nancy and Blanchot." In *Research in Phenomenology* 23:3–21.

Bersani, Leo. 1986. *The Freudian Body: Psychoanalysis and Art.* New York: Columbia University Press.

———. 1990. "Pynchon, Paranoia, and Literature." In *The Culture of Redemption.* Cambridge: Harvard University Press.

Blanchot, Maurice. 1988. *The Unavowable Community*, translated by P. Joris. Berrytown, N.Y.: Station Hill Press.

Blonsky, Marshall, ed. 1985. *On Signs.* Baltimore: Johns Hopkins University Press.

Borges, Jorge Luis. 2000. *The Aleph.* Translated by A. Hurley. London: Penguin.

Botton, Alain de. 1994. *Essays in Love.* London: Picador.

Bright, Susie. 1995. "The Pussyshot: Interview with Andrew Blake." In *Sexwise.* San Francisco: Cleis Press.

Broch, Hermann. 2000. *The Sleepwalkers Trilogy* (*The Romantic, The Anarchist, The Realist*), translated by W. and E. Muir. London: Penguin.

Brockman, John. 1973. *Afterwords: Explorations of the Mystical Limits of Contemporary Reality.* New York: Anchor Press and Doubleday.

Brown, K. 2000. "It's Alive." In *New Scientist.* Vol. 168. No. 2262. October 28: 30–3. Online. Internet. http://www.newscientist.com/article/mg16822624.100.html.

Brown, N. O. 1990. *Love's Body.* Berkeley: University of California.

Burgess, Anthony. 1962. *A Clockwork Orange.* New York: W. W. Norton.

———. 1978. *1985.* New York: Little Brown.

Butler, Judith. 1995. "Contingent Foundations: Feminism and the Question of 'Postmodernism.' In *Feminist Contentions: A Philosophical Exchange*, Linda Nicholson, ed. London and New York: Routledge. 35–57.

Cadava, E., P. Connor, et al., eds. 1991. *Who Comes after the Subject?* New York: Routledge.

Carroll, Lewis. 1992. *Alice in Wonderland: Norton Critical Edition*, edited by Donald J. Gray. New York and London: W. W. Norton.

Certeau, Michel de. 1988. *The Practice of Everyday Life*. Translation by Steven Rendall. Berkeley and Los Angeles: University of California Press.

Cioran, E. M. 1987. *History and Utopia*. New York: Seaver Books.

Clark, Andy. 2003. *Natural-born Cyborgs: Minds, Technologies, and the Future of Human Intelligence*. New York: Oxford University Press.

Cleland, John. 2002. *Fanny Hill: Memoires of a Woman of Pleasure*. London: Wildside.

Clemens, Justin, and Dominic Pettman. 2004. *Avoiding the Subject: Media, Culture and the Object*. Amsterdam: Amsterdam University Press.

Conrad, Joseph. 1968. *The Secret Agent*. London: J. M. Dent & Sons.

Critchley, Simon. 2002. "Calm: On Terence Malick's *Thin Red Line*." In *Film-Philosophy*. Vol 6. No. 38. Online. Internet. http://www.film-philosophy.com/vo16–2002/n48critchley.

Damrosch, David. 2003. *What is World Literature?* Princeton: Princeton University Press.

Debord, Guy. 1977. *Society of the Spectacle*. Detroit. Black & Red.

Deleuze, Gilles. 1986. *Cinema I: The Movement-Image*, translated by H. Tomlinson and B. Habberjam. Minneapolis: University of Minnesota.

———. 1991. "A Philosophical Concept." In Cadava, E., P. Connor, et al., eds. *Who Comes after the Subject?* New York: Routledge.

———. 1992. "Postscript on the Societies of Control." In *October* 59 (Winter): 3–7.

———. 1994. *Difference and Repetition*, translated by P. Patton. New York: Columbia University Press.

———. 1995. "Letter to Serge Daney: Optimism, Pessimism, and Travel." In *Negotiations, 1972–1990*, translated by M. Joughin. New York: Columbia University Press.

———. 2000. *Proust and Signs*, translated by R. Howard. Minneapolis: University of Minnesota Press.

———. 2003. *Cinema 2: The Time-Image*, translated by H. Tomlinson and R. Galeta. Minneapolis: University of Minnesota Press.

Deleuze, Gilles, and F. Guattari. 1999. *A Thousand Plateaus: Capitalism and Schizophrenia*, translated by B. Massumi. London: Athlone.

DeLillo, Don. 1985. *White Noise*. London: Picador.

de Man, Paul. 1979. *Allegories of Reading: Figural Language in Rousseau, Nietzsche, Rilke, and Proust*. New Haven, Conn.: Yale University Press.

Derrida, Jacques. 1994. *Specters of Marx: The State of the Debt, the Work of Mourning, and the New International*, translated by P. Kamuf. New York: Routledge.

Derrida, Jacques, and Bernard Stiegler. 2002. "The Discrete Image." In *Echographies of Television: Filmed Interviews*, translated by J. Bajorek. Cambridge, England: Polity Press.

Dick, Philip K. 1996. *Do Androids Dream of Electric Sheep?* New York: Ballantine.
———. 1993. *The World Jones Made.* New York: Vintage.
Durham, S. 1998. *Phantom Communities: The Simulacrum and the Limits of Postmodernism.* Stanford, Calif.: Stanford University Press.
Eliot. T. S. 1917. *Prufrock, and Other Observations.* London: The Egoist Ltd.
Ellis, Bret Easton. 1991. *American Psycho.* New York, Vintage.
Foucault, Michel. 1972. *The Archaeology of Knowledge*, translated by A. M. Sheridan Smith. New York: Pantheon Books.
———. 1984. "On the Genealogy of Ethics: An Overview of Work in Progress." In *The Foucault Reader, edited by* Paul Rabinow. New York: Pantheon.
Foucault, Michel, L. H. Martin, et al. 1988. *Technologies of the Self: A Seminar with Michel Foucault.* Amherst: University of Massachusetts Press.
Fox, Nicols. 2002. *Against the Machine: The Hidden Luddite Tradition in Literature, Art, and Individual Lives.* Washington: Island Press and Shearwater Books.
Freud, Sigmund. 1953–1974. *The Standard Edition of the Complete Psychological Works*, translated by J. Strachey. London: Hogarth.
Frodeman, R. 2003. *Geo-logic: Breaking Ground between Philosophy and the Earth Sciences.* Albany: State University of New York Press.
Geertz, Clifford. 1983. *Local Knowledge: Further Essays in Interpretive Anthropology.* New York: Basic Books.
Gergen, Kenneth J. 1991. *The Saturated Self: Dilemmas of Identity in Contemporary Life.* New York: Basic Books.
Gibson, William. 1984. *Neuromancer.* New York: Ace Books.
———. 1993. "Disneyland with the Death Penalty." In *Wired.* 1.04. Sep/Oct. Online. Internet. http://www.wired.com/wired/archive/1.04/gibson.html.
———. 1997. *Idoru.* London and New York: Penguin.
———. 1999. *All Tomorrow's Parties.* New York: G. P. Putnam's Sons.
———. 2003. *Pattern Recognition.* New York: G. P. Putnam's Sons.
Girard, René. 1988. *Deceit, Desire and the Novel*, translated by Y. Freccero. Baltimore and London: Johns Hopkins University Press.
Godzich, Wlad. 2000. "The Emergent Moment in Literature." *Tamkang Review* 30:4, 3–13.
Hansen, Mark. 2000. "Becoming as Creative Involution? Contextualizing Deleuze and Guattari's Biophilosophy." *Postmodern Culture* 11:1.
———. 2004. "Affect as Medium, or the 'Digital-Facial-Image.'" In *Journal of Visual Culture.* Vol 3. No. 3: 353–357.
Haraway, Donna. 1991. "A Cyborg Manifesto: Science, Technology, and Socialist-Feminism in the Late Twentieth Century." In *Simians, Cyborgs and Women: The Reinvention of Nature.* New York: Routledge.
Hardt, M., and A. Negri. 2000. *Empire.* Cambridge: Harvard University Press.
Harris, Thomas. 2000. *Hannibal.* London, Arrow.
Hayles, N. Kathryn. 1999. *How We Became Posthuman: Virtual Bodies in Cybernetics, Literature, and Informatics.* Chicago: University of Chicago Press.

Heidegger, Martin. 1969. *The Essence of Reasons*, translated by Terence Malick. Evanston: Northwestern University Press.

———. 1977a. "The Question Concerning Technology." In *The Question Concerning Technology and Other Essays*, translated by W. Lovitt. New York: Harper & Row.

———. 1977b. "The Age of the World Picture." In *The Question Concerning Technology and Other Essays*, translated by W. Lovitt. New York: Harper & Row.

———. 1977c. "The Turning." In *The Question Concerning Technology and Other Essays*, translated by W. Lovitt. New York: Harper & Row.

———. 1977d. "Science and Reflection." In *The Question Concerning Technology and Other Essays*, translated by W. Lovitt. New York: Harper & Row.

———. 1996. *Being and Time*, translated by J. Stambaugh. Albany: State University of New York Press.

———. 1998. "Letter on 'Humanism.'" In *Pathmarks*, edited by W. McNeill, translated by F. A. Capuzzi. Cambridge: Cambridge University Press.

———. 2002. "The Principle of Identity." In *Identity and Difference*, translated by J. Stambaugh. Chicago: University of Chicago Press.

Highfield, Roger. 2001. "Gene Map Shows We're Not So Special," *The Daily Telegraph*. February 13. Online. Internet.

Hoban, Russel. 1980. *Riddley Walker*. New York. Summit Books.

Houllebecq, Michel. 2001. *Atomized*, translated by F. Wynne. London: Vintage.

Hunt, Lynn (ed.). 1996. *The Invention of Pornography: Obscenity and the Origins of Pornography, 1500–1800*. New York: Zone.

Irigaray, Luce. 2000. *To Be Two, translated by* M. M. Rhodes and M. F. Cocito-Monoc. London; New Brunswick, N.J.: Athlone Press.

Jameson, Frederic. 1983. "On Chandler." In *The Poetics of Murder: Detective Fiction and Literary Theory*, edited by G. W. Most and W. W. Stowe. New York: Harcourt Brace Jovanovich.

Jaynes, J. 1976. *The Origin of Consciousness in the Breakdown of the Bicameral Mind*. Boston: Houghton Mifflin.

Judge, Anthony. 2001. "Missiles, Missives, Missions and Memetic Warfare: Navigation of Strategic Interfaces in Multidimensional Knowledge Space." Online. Internet. http://www.laetusinpraesens.org/docs/missile.php

Kendrick, Walter. 1996. *The Secret Museum: Pornography in Modern Culture*. Berkeley: University of California Press.

Kierkegaard, Søren. 1987. "The Seducer's Diary." Part 1 in *Either/Or*. H. V. a. E. H. H. Hong. Princeton, N.J.: Princeton University Press.

Kilgour, M. 1990. *From Communion to Cannibalism: An Anatomy of Metaphors of Incorporation*. Princeton, N.J.: Princeton University Press.

King James Bible. Online. Internet. University of Michigan. http://www.hti.umich.edu/k/kjv/.

Kipnis, Laura. 2003. *Against Love: A Polemic*. New York: Pantheon.

Kittler, Friedrich A. 1990. *Discourse Networks 1800/1900*, translated by M. Metteer and C. Cullens. Stanford, Calif.: Stanford University Press.

———. 1999. *Gramophone, Film, Typewriter*, translated by Geoffrey Winthrop-Young and Michael Wutz. Stanford: Stanford University Press.

Kunhardt, P. B. e. a. 1995. *P. T. Barnum: America's Greatest Showman*. New York: Alfred A. Knopf.

Kunstler, James Howard. 1993. *The Geography of Nowhere: The Rise and Decline of America's Man-Made Landscape*. New York: Simon & Schuster.

Lacan, Jacques. 1977. *Écrits: A Selection*. New York: Norton.

———. 1978. *The Four Fundamental Concepts of Psycho-analysis*. New York: Norton.

———. 1999. *On Feminine Sexuality: The Limits of Love and Knowledge, 1972–1973*, edited by J.-A. Miller, translated by B. Fink. New York & London: W. W. Norton.

Lanchester, John. 1996. *The Debt to Pleasure*. London: Picador.

Lawrence, D. H. 1994. *Lady Chatterley's Lover*. London: Penguin.

Levinas, Emmanuel. 1985. *Ethics and Infinity: Conversations with Philippe Nemo*, translated by R. A. Cohen. Pittsburgh: Duquesne University Press.

Lévy, Pierre. 1998. *Becoming Virtual: Reality in the Digital Age*, translated by R. Bononno. New York and London: Plenum.

Lovink, Geert. 2005. "The Principle of Notworking: Concepts in Critical Internet Culture." Online. Internet. http://www.hva.nl/lectoraten/0109–050224-lovink.pdf.

Lucretius. 1994. *On the Nature of the Universe*, translated by R. E. Latham. London: Penguin.

Luhmann, Niklas. 1996. "Complexity, Structural Contingency and Value Conflicts." In *Detraditionalization: Critical Reflections on Authority and Identity*, edited by P. Heelas, S. Lash, and P. Morris. London: Blackwells.

———. 1998. *Love as Passion: The Codification of Intimacy*, translated by J. Gaines and D. L. Jones. Stanford, Calif.: Stanford University Press.

Luke, Timothy W. 1999. "Simulated Sovereignty, Telematic Territoriality: The Political Economy of Cyberspace." In *Spaces of Culture: City, Nation, World*, edited by M. Featherstone and S. Lash. London: Sage.

Lyon, David. 1994. "From Big Brother to Electronic Panopticon." In *The Electronic Eye: The Rise of Surveillance Society*. Minneapolis: University of Minnesota Press.

Maar, Michael. 2005. *The Two Lolitas*, translated by Perry Anderson. New York and London: Verso.

Maffesoli, Michel. 1993. *The Shadow of Dionysus: A Contribution to the Sociology of the Orgy*, translated by C. Linse and M. K. Palmquist. Albany: State University of New York Press.

———. 1996. *The Time of the Tribes: The Decline of Individualism in Mass Society*, translated by D. Smith. London: Sage.

Maines, R. P. 1999. *The Technology of Orgasm: "Hysteria," the Vibrator, and Women's Sexual Satisfaction.* Baltimore: Johns Hopkins University Press.
Marrati, Paola. 2001. "Against the Doxa: Politics of Immanence and Becoming-Minoritarian." In *Micropolitics of Media Culture: Reading the Rhizomes of Deleuze and Guattari.* Patricia Pisters and Catherine M. Lord, eds. Amsterdam: University of Amsterdam Press.
Marinetti, F. T. 1989. *The Futurist Cookbook*, translated by Susan Brill. San Francisco: Bedford Arts.
Marx, Karl. 1977. *Capital.* Vol. 1, translated by B. Fowkes. New York: Vintage.
Mazlish, Bruce. 1993. *The Fourth Discontinuity: The Co-Evolution of Humans and Machines.* New Haven: Yale University Press.
McLuhan, Marshall, and Fiore, Q. 1997. *War and Peace in the Global Village*, edited by J. Agel. San Francisco: Hardwired.
McNair, Brian. 1996. *Mediated Sex: Pornography and Postmodern Culture.* London and New York: Arnold.
Miller, Walter M. 1960. *A Canticle for Leibowitz.* Philadelphia: Lippincott.
Mitchell, W. J. T. 2002. "Showing Seeing: a Critique of Visual Culture." In *Journal of Visual Culture*, Vol. 1. No. 2.: 165–181.
Montaigne. 1970. *Essays*, translated by J. M. Cohen. London: Penguin.
Mori, Masahiro. 1970. "Bukimi no tani" (The Uncanny Valley). In *Energy*. 7(4): 33–35.
Morris, Paul. 1996. "Community beyond Tradition." In *Detraditionalization*, edited by P. Heelas et al. London: Blackwells.
Murakami, Haruki. 1993. *Hard-Boiled Wonderland and the End of the World*, translated by A. Birnbaum. New York: Vintage.
———. 2000. *Norwegian Wood*, translated by Jay Rubin. New York: Vintage.
———. 2002. *A Wild Sheep Chase*, translated by A. Birnbaum. New York: Vintage.
Musil, Robert. 1996. *The Man without Qualities*, translated by S. Wilkins. New York: Vintage.
Nabokov, Vladimir. 1991. *The Annotated Lolita*, edited by A. Appel, Jnr. New York: Vintage.
Nancy, Jean-Luc. 2000. *Being Singular Plural*, translated by R. D. Richardson and A. E. O'Byrne. Stanford, Calif.: Stanford University Press.
———. 1991. *The Inoperative Community*, translated by P. Connor et al. Minneapolis: University of Minnesota Press.
Negroponte, Nicholas. 1995. *Being Digital.* New York: Knopf.
Nelson, Victoria. 2001. *The Secret Life of Puppets.* Cambridge: Harvard University Press.
Nichols, Bill. 1996. "The Work of Culture in the Age of Cybernetic Systems." In *Electronic Culture, edited by* T. Druckrey. New York: Aperture Press.
Nietzsche, Friedrich. 1956. *The Genealogy of Morals*, translated by F. Golffing. New York: Anchor Books, Doubleday.

———. 1992. *Ecce Homo: How One Becomes What One Is*, translated by R. J. Hollingdale. London and New York: Penguin Classics.
———. 1995. *Human, All Too Human*, translated by G. Handwerk. Stanford, Calif.: Stanford University Press.
———. 2001. *The Gay Science*, edited by B. Williams, translated by J. Nauckhoff. Cambridge: Cambridge University Press.
Observer, The. 2000. "Quotes of the Year: Did They Really Say That?" December 31, 32.
Odell, David. 2001. *A Rushed Quality: Observations on Living. Blackjelly* (January), http://www.blackjelly.com/Mag2/features/rushed.htm.
Oxford English Dictionary. 2006. Online. Internet. http://www.oed.com.
Pesic, Peter. 2002. *Seeing Double: Shared Identities in Physics, Philosophy, and Literature*. Cambridge: MIT Press.
Pettman, Dominic. 2002. *After the Orgy: Toward a Politics of Exhaustion*. Albany: State University of New York Press.
Piane, Renee. 2001. *Love Mechanics: Power Tools to Build Successful Relationships with Women*. Santa Monica: Love Works Publishing.
Plato. 1999. "The Symposium." In *The Essential Plato*, translated by B. Jowett and M. J. Knight. New York: Quality Paperback.
———. (1999b). "Phaedrus." In *The Essential Plato*, translated by B. Jowett and M. J. Knight. New York: Quality Paperback.
———. (1999c). "The Republic." In *The Essential Plato*, translated by B. Jowett and M. J. Knight. New York: Quality Paperback.
Polhemus, Robert M. 1990. *Erotic Faith: Being in Love from Jane Austen to D. H. Lawrence*. Chicago: University of Chicago Press.
Poster, Mark. 1996. "Databases as Discourse, or Electronic Interpellations." In *Detraditionalization*, edited by P. Heelas et al. London: Blackwells.
Praz, Mario. 1960. *The Romantic Agony*. Glasgow: Fontana.
Proust, Marcel. 1989. *Remembrance of Things Past*, translated by C. K. S. Moncrieff and T. Kilmartin. London: Penguin.
Pynchon, Thomas. 1966. *The Crying of Lot 49*. Philadelphia and New York: J. B. Lippincott.
———. 1981. *Gravity's Rainbow*. London: Picador.
———. 1998. *Mason & Dixon*. London: Vintage.
Reuters. 2002. "Scientists Plan To Shake Hands Over Internet." In *USA Today*. October 10. Online. Internet. http://www.usatoday.com/tech/news/tech innovations/2002–10–29-handshake1_x.htm.
Robins, Kevin. 2000. "Cyberspace and the World We Live In." In *The Cybercultures Reader*, edited by D. Bell and B. M. Kennedy. London and New York: Routledge.
Robnik, Drehli. 2002. "Saving One Life: Spielberg's *Artificial Intelligence* as Redemptive Memory of Things." In *Jumpcut: A Review of Contemporary Cinema*. Issue 45. Online. Internet. http://www.ejumpcut.org/archive/jc45.2002/robnik/AItext.html.

Ross, Andrew. 1999. *Celebration Chronicles: Life, Liberty, and the Pursuit of Property Value in Disney's New Town.* New York: Ballantine.
Ross, Daniel. 2006. "Democracy, Authority, Narcissism: From Agamben to Stiegler." In *Contretemps* 6 (January): 74–85. http://www.usyd.edu.au/contretemps/6January2006/ross.pdf.
Roth, Philip. 2001. *The Dying Animal.* New York: Vintage International.
Rougement, D. 1983. *Love in the Western World.* Princeton, N.J.: Princeton University Press.
Rousseau, Jean-Jacques. 1945. *The Confessions.* New York: Modern Library.
Said, E. W. 1978. *Orientalism.* New York: Pantheon Books.
Sartre, Jean-Paul. 1956. "Concrete Relations with Others." In *Being and Nothingness: An Essay on Phenomenological Ontology*, translated by Hazel E. Barnes. New York: Philosophical Library.
Sarup, Madan. 1992. *Jacques Lacan.* London: Prentice-Hall.
Schreck, Gina. 2002. *The Marriage Mechanics: A Tune Up for the Highway of Love.* Talking Fish Publishers.
Sebald, W. G. 1997. *The Emigrants*, translated by M. Hulse. London: Harvill Press.
Shaviro, Steven. 2001. "Super Transgressions: The New Pathologies." In *The Stranger.* (Nov 29—Dec 5). Online. Internet. http://www.thestranger.com/seattle/Content?oid=9420
———. 2002. "The Erotic Life of Machines." In *Parallax* 8, no. 4 (October-December) 21–31.
———. 2003. *Connected: What It Means to Live in the Network Society.* Minneapolis: University of Minnesota Press.
———. 2004. "Returning to the Scene of the Perfect Crime, Or, How I Learned to Stop Worrying and Love the Virtual." Unpublished Paper.
Shohat, Ella, and Robert Stam. 2002. *Unthinking Eurocentrism.* London & New York: Routledge.
Shu Lea Cheang. 2001. Interview with Geert Lovink. Online. Internet. http://www.othlo.com/haudiovisuales/eventos/01mediarama01/01femart/05shu.htm.
———. 2002. "I.K.U. The Movie," Promotional Material. Online. Internet. http://www.i-k-u.com.
Silverman, Kaja. 2000. *World Spectators.* Stanford, Calif.: Stanford University Press.
Sloterdijk. Peter. 2001. *The Elmauer Rede: Rules for the Human Zoo: A Response to the Letter on Humanism*, translated by M. V. Rorty. Online. Internet.
Soler, Colette. 2000. "The Curse on Sex." In Renata Salecl (ed.). *Sexuation.* Durham: Duke University Press.
Spinoza, Benedictus de. *Ethics.* 2000. Oxford: Oxford University Press.
Stendhal. 1975. *Love*, translated by G. and S. Sale. London: Penguin.

Stiegler, Bernard. 1998. *Technics and Time*. Vol. 1, *The Fault of Epimetheus*, translated by R. Beardsworth and G. Collins. Stanford, Calif.: Stanford University Press.

Taubin, A. 1999. "Imperfect Love." *Film Comment* 35(5):24.

Teller, Juergen. 1999. *Juergen Teller Go-Sees*. Berlin: Scalo Verlag.

———. 2001. *Tracht*. New York: Lehmann Maupin Gallery.

Terada, Rei. 2001. *Feeling in Theory: Emotion After the "Death of the Subject."* Cambridge: Harvard University Press.

Trinh, T. M.-H. 1989. *Woman, Native, Other: Writing Postcoloniality and Feminism*. Bloomington: Indiana University Press.

Vonnegut, Kurt. 1991. *Slaughterhouse Five; or. The Children's Crusade, A Duty Dance with Death*. New York: Dell.

Wall, Thomas Carl. 1999. *Radical Passivity: Levinas, Blanchot, and Agamben*. New York: State University of New York Press.

Waugh, Evelyn. 1943. *Decline and Fall*. Boston: Little Brown.

Weber, S. 1996. *Mass Mediaurus: Form, Technics, Media*, edited by A. Cholodenko. Stanford, Calif.: Stanford University Press.

Weiss, A. S. 1989. *The Aesthetics of Excess*. Albany: State University of New York Press.

Williams, Linda. 1990. *Hard Core: Power, Pleasure, and the "Frenzy of the Visible."* Berkeley: University of California Press.

Woolf, Virginia. 1979. *The Waves*. New York and London: Harcourt.

Zandonella, Catherine. 2001. "Spinning a Yarn for Miniature Electronics." In *New Scientist* 2289 (May 5): 20. Online. Internet. http://www.newscientist.com/article/mg17022892.900.html.

Zizek, Slavoj. 1997. *The Plague of Fantasies*. London and New York: Verso.

———. 1999. *The Ticklish Subject: The Absent Centre of Political Ontology*. London and New York: Verso.

———. 2001. *Did Somebody Say Totalitarianism?: Five Interventions in the (Mis)-use of a Notion*. London and New York: Verso.

VISUAL RESOURCES

Barison, David, and Daniel Ross. 2004. *The Ister*.

Bergman, Ingmar. 1953. *Monika*.

Blake, Andrew. 1989. *Night Trips*.

———. 1990. *House of Dreams*.

Brain Story: The Mind's Eye. 2000. BBC television.

Frankenheimer, John. 1966. *Seconds*.

Hill, Tim. 1999. *Muppets from Space*.

Hitchcock, Alfred. 1958. *Vertigo*.

Jones, Terry. 1979. *Life of Brian*.

Jonze, Spike. 1999. *Being John Malkovich*.

Judge, Mike. 1999. *Office Space.*
Just Shoot Me! 1997–2003. NBC television.
King, Zalman. 1990. *Wild Orchid.*
Koreeda, Hirokazu. 1998. *Wandāfuru Raifu (After Life).*
Kubrick, Stanley. 1962. *Lolita.*
———. 1968. *2001: A Space Odyssey.*
———. 1971. *A Clockwork Orange.*
———. 1987. *Full Metal Jacket.*
———. 1999. *Eyes Wide Shut.*
Luhrmann, Baz. 1996. *Romeo + Juliet.*
Lyne, Adrian. 1986. *9½ Weeks.*
Malick, Terence. 1998. *The Thin Red Line.*
Marker, Chris. 1962. *Le Jetée.*
Miller, George. 1981. *Mad Max 2.*
Mir Mortals. 1998. BBC Horizon.
Polanski, Roman. 1992. *Bitter Moon.*
Raimi, Sam. 2002. *Spider-Man.*
Renders, Pierre-Paul. 2000. *Thomas est amoureux (Thomas is in Love).*
Rohmer, Eric. 1972. *L'amour l'après-midi (Chloe in the Afternoon).*
Sakaguchi, Hironobu. 2001. *Final Fantasy: The Spirits Within.*
Shu Lea Cheang. 2000. *I.K.U.*
Simpsons, The. 1989–2006. Fox television.
Soderbergh, Steven. 1989. *Sex, Lies, and Videotape.*
Spielberg, Steven. 2001. *Artificial Intelligence: AI.*
Tykwer, Tom. 1998. *Lola rennt (Run Lola Run).*
Vertov, Dziga. 1929. *Chelovek s kino-apparatom (The Man With a Movie Camera).*
Wachowski, Andy and Larry Wachowski. 1999. *The Matrix.*
———. 2003. *The Matrix Reloaded.*
———. 2003. *The Matrix Revolutions.*
Wong Kar-wai. 1994. *Chung hing sam lam (Chungking Express).*
———. 1995. *Duo luo tian shi (Fallen Angels).*
———. 2000. *Fa yeung nin wa (In The Mood For Love).*
Woo, John. 1997. *Face/Off.*

AUDIO RESOURCES

Brown, Sam. 1993. "Fear of Life." *43 Minutes.* LP CD. Pod.
Cat Power. 1998. "Colors and the Kids." *Moon Pix.* LP CD. Matador.
Crosby, Stills, Nash and Young. 1971. "Triad." *Four Way Street.* LP record. WEA / Atlantic.
Gaynor, Janet. 1929. "If I Had a Talking Picture of You." From the film *Sunny Side Up.* Directed by David Butler.

Lynch, Liam. 2002. "United States of Whatever." CD Single. Global Warming.
Madonna. 1990. "Vogue." CD Single. Sire/Rhino.
Nirvana. 1991. "Smells Like Teen Spirit." *Nevermind.* LP CD. Geffen.
Pink. 2002. "Don't Let Me Get Me." CD Single. BMG.

INDEX

Lightning Source UK Ltd.
Milton Keynes UK
UKHW03n0225070418
320635UK00001B/2/P